12-6-76

11323
1-05002

Dictyostelium discoideum

A Developmental System

Dictyostelium discoideum

A Developmental System

William F. Loomis

Department of Biology
University of California at San Diego
La Jolla, California

ACADEMIC PRESS   New York   San Francisco   London   1975

A Subsidiary of Harcourt Brace Jovanovich, Publishers

ACADEMIC PRESS, INC.
111 Fifth Avenue, New York, New York 10003

United Kingdom Edition published by
ACADEMIC PRESS, INC. (LONDON) LTD.
24/28 Oval Road, London NW1

Library of Congress Cataloging in Publication Data

Loomis, William F
    Dictyostelium discoideum.

    Bibliography: p.
    1. Developmental biology.   2.   Dictyostelium discoi-
deum.   I.   Title.   [DNLM: 1.   Myxomycetes.   QW180 L863d]
QP83.8.L66   593'.117   74-27786
    ISBN 0–12–456150–0

QH491
.L86

Contents

Preface

Dictyostelium discoideum has been attracting attention from an increasing number of biologists who have become aware of its unique advantages as a model eukaryotic system in which to study a variety of developmental problems. The organism lies in the transition zone between major phylogenetic categories as shown by the fact that many botanists consider it a lower plant while many zoologists consider it a lower animal. It has many of the attributes and experimental advantages of microorganisms but displays multicellular phenomena similar to those of metazoans. Its rapid life cycle and small size make it convenient for cytological studies, while its synchronous development allows biochemical analyses.

The genetic and biochemical studies which, in recent years, have begun to elucidate the physiological processes that underlie morphological phenomena form the focus of this review. Earlier work has been summarized in the excellent book of J. T. Bonner, "The Cellular Slime Molds." Although many of the present studies do not provide conclusive answers to the developmental problems, they better delineate the questions and indicate fruitful approaches which can be pursued.

Biochemical processes which evolved several billion years ago in the simplest unicellular organisms have been conserved down to present day higher organisms to a surprising extent. Underneath the variety of

ix

morphological specializations of multicellular organisms, the pathways of energy production and biosynthesis have remained almost unchanged. The universality of the genetic code and the unique mechanism of protein synthesis attest to the fact that once the necessary degree of precision of gene expression was attained few further modifications evolved.

Higher organisms came into existence about a billion years ago and seem to have followed a pathway to specialization which often involved using preexisting physiological functions in new combinations rather than the evolution of completely new ones. Thus, a basic understanding of a process in one organism often has much to say about related processes in other systems.

I have attempted to put in focus most of the major studies on *D. discoideum*. I have not tried to include all of the amazingly diverse facts concerning this organism which have appeared in recent years. Many preliminary observations cannot yet be incorporated into a coherent fabric, and rather than attempting to be exhaustively complete, I have tried to give the present approaches in a concise and useful form. For the sake of continuity and brevity, studies on related slime molds have not been included, although some species of the Acrasiales are described for the purpose of comparison in Chapter 1.

The recent interest in development of *D. discoideum* has generated a greatly increased flow of published works. Many of these repeat and extend earlier work. Where possible the original papers have been referred to in the text; however, it seemed cumbersome in this type of book to exhaustively reference each observation discussed. Instead, pertinent publications are indicated so that the reader can gain further information on a point of interest. Excellent reviews of various aspects of the development of *D. discoideum* have recently been published and provide more complete referencing (see Garrod and Ashworth, 1973; Newell, 1971; Bonner, 1971; Killick and Wright, 1974; Gregg and Badman, 1973; and Sussman and Sussman, 1969). An effort has been made to make the Bibliography on *D. discoideum* as complete as possible to 1974.

I am indebted to my colleagues, Randall Dimond, Paul Farnsworth, Richard Firtel, Steven Free, Hudson Freeze, and Ken Poff, for careful reading of the manuscript and many helpful discussions. I am particularly indebted to Dr. Paul Farnsworth for providing the illustrations for this book. Studies in my laboratory have been supported by grants from the National Science Foundation and the National Institutes of Health.

William F. Loomis

Chapter 1

The Life Cycle

An Overview

Dictyostelium discoideum is found in nature as a soil amoeba in forest detritus. It was first discovered by Raper (1935) in the woods of North Carolina but has since been found in similar ecological environments throughout the world (Cavender and Raper, 1965a,b,c, 1968). The cells feed on the bacteria of decaying fallen leaves and other matter and divide by binary fission much like any other soil amoebae. However, when the local environment is depleted of food source, a unique and characteristic series of developmental events occurs which distinguishes *D. discoideum* from less social organisms. The cells collect in large streaming patterns to form groups containing up to 10^5 cells (Fig. 1.1). These aggregates become integrated by deposition of a surface sheath covering the whole mound. The aggregate topples over onto the surface and migrates horizontally still surrounded by the sheath, and as such is referred to as a pseudoplasmodium, grex, or slug (Fig. 1.2).

After a variable period of migration, the pseudoplasmodium stops

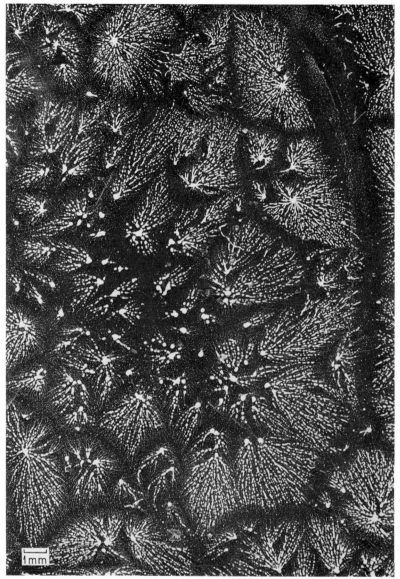

Fig. 1.1. Aggregation patterns. A shadowgraph of many cells aggregating to centers. Each individual amoeba appears as a small white dot, while aggregation centers appear as clear areas.

and the anterior tip moves vertically in preparation for the terminal differentiations which form the fruiting body. At this point cells near the tip of the mass begin to synthesize increased amounts of cellulose

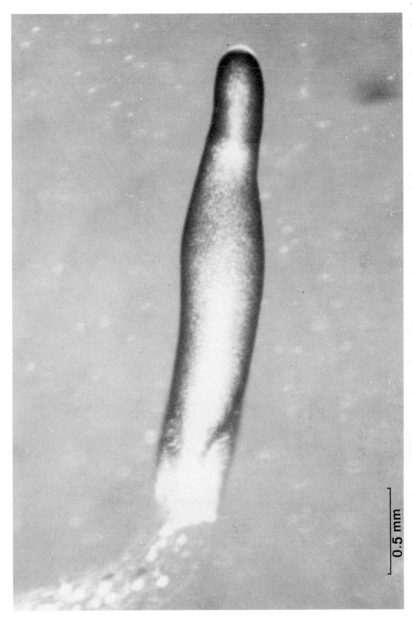

Fig. 1.2. A migrating pseudoplasmodium seen from above. The tapered anterior tip and the collapsed tube of surface sheath left behind are clearly visible.

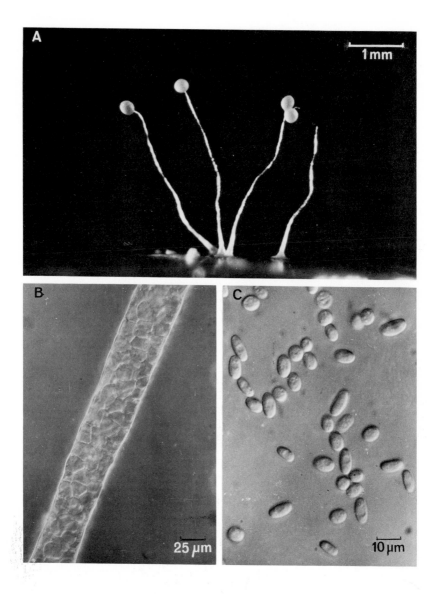

Fig. 1.3. Mature fruiting bodies. (A) Sori supported on tapering stalks. (B) Differentiated stalk cells (Nomarski optics); angular cellulose walls can be seen surrounding the cells, which are enclosed in the stalk sheath. (C) Differentiated spores (Nomarski optics).

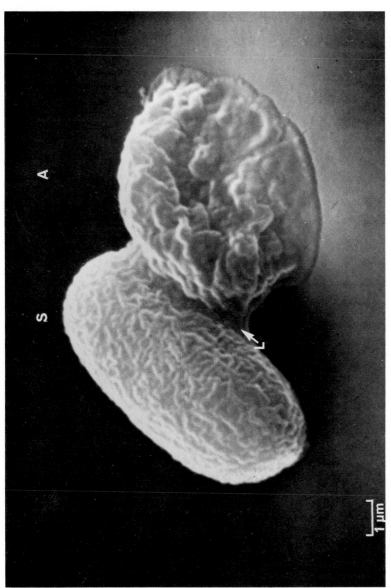

Fig. 1.4. Germinating spore. The emerging amoeba (A) can be seen escaping the spore case (S) through a longitudinal slit (L). The specimen was fixed in 2% glutaraldehyde, postfixed in 1% osmium tetroxide and critical point dried from Freon (SEM).

forming a sheath within which the cells expand and vacuolize to form the stalk. Once the stalk sheath makes contact with the underlying support all further expansion results in extending the stalk vertically. The peripheral cells are lifted up on the rising stalk which engulfs more and more cells at its open upper end. The process has been compared to a fountain running backward (Bonner, 1967). Once inside the stalk, the cells expand and vacuolize lifting the whole mass up further. As the mass rises, the peripheral cells begin to encapsulate and form the apical sorus. When all of the cells have either been engulfed into the stalk or have encapsulated to form spores, the formation of the fruiting body is complete (Fig. 1.3).

Stalk cells are no longer viable after vacuolization but the spores can remain viable through extended periods of dehydration, starvation, and elevated temperature. When the spores are dispersed, they germinate by splitting the spore case longitudinally and escaping as small but normal amoebae (Fig. 1.4). The sequence of morphological stages is shown schematically in Fig. 1.5.

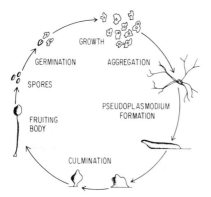

Fig. 1.5. Life cycle of *D. discoideum.* Up to 10^5 amoebae aggregate and form pseudoplasmodia. Fruiting bodies formed from such aggregates are about 1–2 mm high and contain about 7×10^4 ellipsoid spores.

The apparent selective advantage of this complex series of stages is 2-fold: to permit the dispersal of the cells from an area in which they were starving and to provide a dormant stage to resist unfavorable conditions. Impairment of either function would most likely be detrimental to the cells' survival.

Synchronous development of up to 10^{11} cells can be achieved in the laboratory by washing amoebae free of exogenous nutrients and depositing them on filter papers soaked in a buffered salt solution (Sussman, 1966a). Under these conditions the cells proceed through the morphological stages in 24 hours with greater than 95% synchrony. The developmental stages are shown in Figs. 1.6 and 1.7.

Related Species

Dictyostelium discoideum is a member of the class Acrasieae which includes those species of free-living amoebae that lack a flagellated stage and aggregate to form fruiting bodies. Bonner (1967) has suggested that the Acrasiales can be divided into two, possibly unrelated families, the Guttulinaceae and the Dictyosteliaceae (Fig. 1.8).

The Dictyosteliaceae include several clearly defined species. The small amoebae in these species exhibit a number of filopods and stream chemotactically to form aggregates. The size of the aggregates varies considerably in each species; however, fruiting bodies of *Dictyostelium lacteum* and *Dictyostelium minutum* are never more than a tenth the size of those of *D. discoideum*. *Dictyostelium lacteum* is further characterized by forming spherical rather than ellipsoid spores. Aggregation in *D. minutum* has been found to be less complex than in other species (Raper, 1960a). Development in these species follows much the same pattern as occurs in *D. discoideum*. In the larger species, *Dictyostelium polycephalum*, the cells aggregate and form pseudoplasmodia much as do those of *D. discoideum;* however, after the migrating stage, the pseudoplasmodium rounds up into a globular mass from which up to 9 papillae or tips emerge, each of which initiates stalk formation and results in a fruiting body. As these culminate, the rising stalks become cemented together for most of their length. The final form resembles a group of *D. discoideum* fruiting bodies held in a narrow vase (Raper, 1956b).

There are two other species of the genus *Dictyostelium*, namely, *D. mucoroides* and *D. purpureum*. These differ from each other in that the latter accumulates a purple pigment in the spore mass. Aggregation in these species is well developed and occurs in large wheellike patterns. The pseudoplasmodia are similar in size and shape to those of *D. discoideum*. However, stalk formation is initiated as soon as the pseudoplasmodia begin to migrate over the surface. The stalk is formed near the anterior end where cells are drawn into it and proceed to vacuolize. The stalk then passes down the length of the pseudoplasmodium and out the back. The stalk is sufficiently strong to support the full weight of the pseudoplasmodium and thus allows aerial migration over obstacles which would bar the advance of *D. discoideum*. During prolonged migration a considerable proportion of the pseudoplasmodium may be expended in stalk formation. Migration in these species is terminated when spore differentiation takes place and the remaining prestalk cells are used up in stalk formation. In *D. mucoroides* and *D. purpureum* the final structure closely resembles that of *D. discoideum*, except that they lack a basal disc.

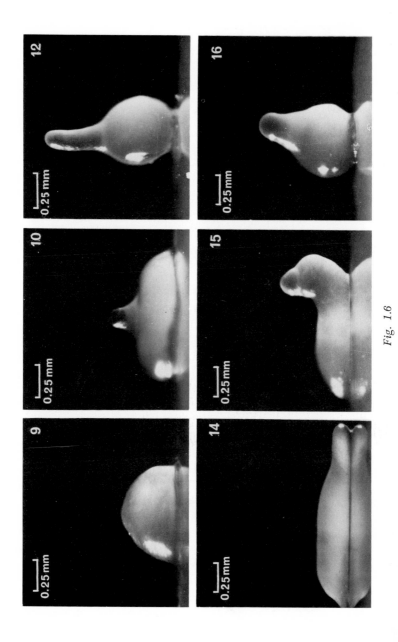

Fig. 1.6

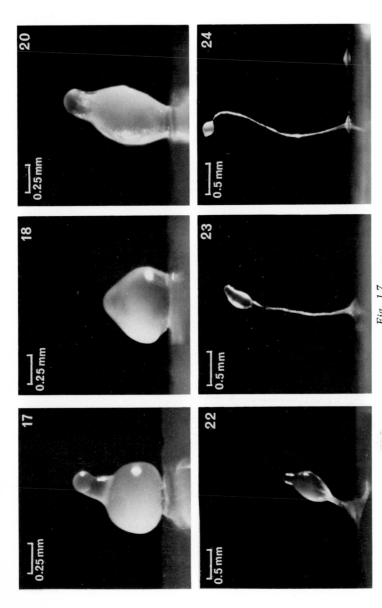

Fig. 1.7

Figs. 1.6 and 1.7. Developmental stages. The sequence of morphological changes observed during normal development of *D. discoideum* NC-4 incubated at 22°C on 2% agar. The numbers indicate the approximate time in hours after the initiation of development. When development occurs on filters supported on pads soaked in 10^{-3} *M* phosphate buffer pH 6.5 containing 9 mg/ml NaCl and 0.5 mg/ml streptomycin sulfate development is highly synchronous; however, the cells remain in the form seen at 12 for about 5 hours and then proceed to the form seen at 17. When cells develop on agar, they can be forced to remain in the form seen at 14 for extended periods.

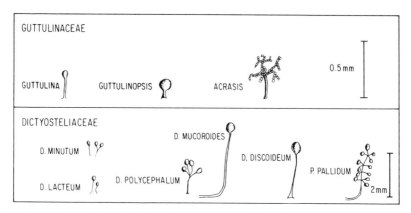

Fig. 1.8. Mature fruiting bodies of some species of the Acrasiales. A more complete description of these species can be found in Bonner (1967).

The pattern of early development is similar in the closely related genus *Polysphondylium*. This genus includes two species, *P. pallidum* and *P. violaceum*, which differ from each other only in that fruiting bodies of the latter are somewhat larger and accumulate a violet pigment in the spore mass (Raper, 1960a). Shaffer (1961b) has shown that aggregation in these species is initiated by specialized "founder" cells which attract surrounding amoebae. Among the *Dictyostelium* species only *D. minutum* seems to aggregate in this manner. Migration in *Polysphondylium* proceeds with concomitant stalk formation in the same fashion as migration in *D. mucoroides*. However, culmination in *Polysphondylium* results in a very different terminal morphology of the fruiting body. Instead of a single apical spore mass, each stalk of *Polysphondylium* carries 8 to 10 whorls, fairly evenly spaced along its length. As the mass of cells is lifted off the support in preparation for culmination, groups of cells are left behind at roughly hourly intervals. These rings of cells subdivide into several small groups of cells and each group produces a short stalk and apical ball of spores. The final form has been likened to that of a diminutive pine tree (Raper, 1960a). A variety of other species of the Dictyosteliaceae have been isolated but all differ considerably from *D. discoideum*. They are well described in Bonner's (1967) book, "The Cellular Slime Molds."

Many of the differences which characterize the above mentioned species are readily mutable characteristics, such as spore color or fruiting body size. This raises the possibility that many of the species are simply variants which recently diverged from a common ancestral form. Few differences would then be expected to be found in the detailed biochemical processes which direct development in the species. Other differences, such as branched versus unbranched culmination, may suggest basic differences in adaptive behavior.

Recent molecular and cytological studies have concentrated on *D. discoideum* to a far greater extent than on related species. This choice has been partially driven by the momentum of the previous body of work on the species, and partially reflects the ease with which the developmental forms can be manipulated since they are large enough to be seen clearly and do not have a rigid stalk during the migration phase. Conditions have been developed for convenient growth of this species in the laboratory and for synchronous development of a sufficient number of cells for biochemical analysis at every stage.

Chapter 2

Amoeboid Stage

Growth Conditions

Dictyostelium discoideum can be grown in the laboratory on a wide variety of bacteria (Raper, 1937). However, *Escherichia coli* B/r or *Klebsiella aerogenes* are most frequently used as the bacterial associate. Suspensions of bacteria, either live or after autoclaving, support rapid growth of the amoebae in liquid (Raper, 1939). For routine growth it is more convenient to grow the amoebae in association with growing bacteria on a solid medium such as the standard one, "SM," containing 10 gm dextrose, 10 gm bactopeptone, 1 gm yeast extract, 1 gm $MgSO_4$, and 10^{-3} M phosphate buffer pH 6.4 per liter of 2% agar (Sussman, 1966). Plates containing this medium can be inoculated with about 10^8 bacteria and 10^5 amoebae to achieve maximum yield of *D. discoideum* (about $4-5 \times 10^8$ amoebae per plate). Initially the bacteria overgrow the plate forming a lawn. Bacterial growth then slows while the amoebae continue to grow with a doubling time of about 4 hours until they have ingested most of the bacteria and have cleared the plate. The amoebae can be washed off the surface and separated from residual bacteria by differential centrifugation.

Although it has been impossible to grow the wild-type isolates in a defined medium in the absence of bacteria, several derivatives of *D. discoideum* NC-4 have been selected for axenic growth in liquid medium. R. Sussman and Sussman (1967) first isolated a strain, Ax-1, which could grow in a medium containing proteose peptone, dextrose, yeast extract, a liver extract, and fetal calf serum. Subsequently, several other strains were selected for growth in a simplified medium (HL-5) containing only 10 gm proteose peptone, 5 gm yeast extract, and 10 gm dextrose per liter. Strain Ax-2 has been used in several laboratories in Europe and its growth has been extensively characterized (Watts and Ashworth, 1970; Ashworth and Watts, 1970). Strain A-3, which develops somewhat more synchronously than strain Ax-2, has been the subject of extensive developmental studies in the United States (Loomis, 1971; Dimond *et al.*, 1973; Firtel and Bonner, 1972a,b; Rossomando and Sussman, 1973). There are no striking differences between these two strains, and all characteristics found in one that have been tested for in the other have been confirmed. The axenic strains grow exponentially with a doubling

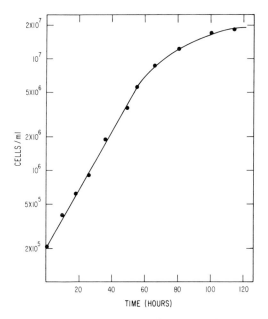

Fig. 2.1. Growth of strain A3 in broth medium. Amoebae were inoculated in 2 liters of HL-5 medium and incubated at 22°C with rapid swirling in a 6-liter Erlenmeyer flask. Cell counts were made on a Coulter Counter Model B.

time of 8–10 hours to a titer of 2×10^6 cells/ml and then at a decreasing rate to a titer of 2×10^7 cells/ml (Fig. 2.1). Two liters of medium incubated with rapid swirling in a 6-liter flask can provide 5×10^9 exponentially growing cells (10 gm dry weight of cells).

Attempts to define a minimal medium have thus far met with little success for any of the axenic strains. It has been possible to replace the requirement for yeast extract with a mixture of amino acids and vitamins, but high molecular weight components of proteose peptone appear to be essential nutrients. Development of a minimal medium will open up the possibility of isolation of well-defined nutritional auxotrophs useful in genetic analyses.

Synchronous growth in HL-5 medium can be achieved with a temperature-sensitive derivative of strain A3 which grows well at 22°C, but is arrested in the G_2 phase at 27°C (Katz and Bourgnignon, 1974). After incubating cultures of this strain for 16 hours at 27°C, growth at 22°C is synchronous for several generations. About 2 hours after cell division, DNA synthesis occurs and proceeds for about 1–2 hours. The S phase is followed by a second growth phase (G_2) of 4 hours. Mitosis and cytokinesis then occur and appear to take 1–2 hours. The cell cycle is shown in Fig. 2.2.

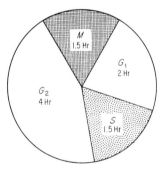

Fig. 2.2. Cell cycle of *D. Discoideum*. Mitosis (*M*) and cell division occur over a period of 1.5 hours. A growth phase (G_1) precedes the period of DNA synthesis (*S*). The second growth phase (G_2) is about 4 hours in duration. The cycle was determined in a derivative of strain A3 growing axenically at 22°C in broth medium (HL-5) (Katz and Bourguignon, 1974).

Amoeboid Movement in *D. discoideum*

On a solid support the amoebae move around by a mechanism of amoeboid movement which appears similar to that of other locomotory eukaryotic cells (Shaffer, 1964a). Thin filopodia are extended a consider-

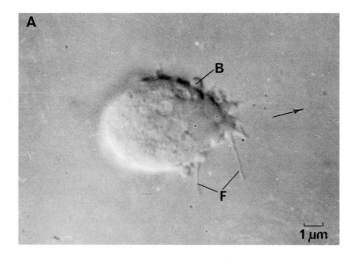

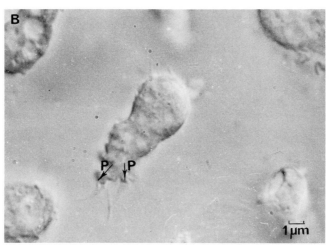

Fig. 2.3. Amoebae. (A) Blebbing of the anterior surface (BL) and projection of filopodia (F) occur in the direction of locomotion indicated by the arrow (Nomarski optics). (B) Pseudopodia (P) can be seen resulting from cytoplasmic flow into filopodia. The arrows indicate the direction of movement (Nomarski optics).

able distance from the cell body, and when these contact a solid support, they may increase in size until they appear as pseudopodia (Fig. 2.3).

Although several theories are presently being considered to account for amoeboid movement in a variety of organisms, a considerable amount of evidence has suggested that cell movement in nonmuscle systems utilizes an actomyosin contractile system similar to that utilized in stri-

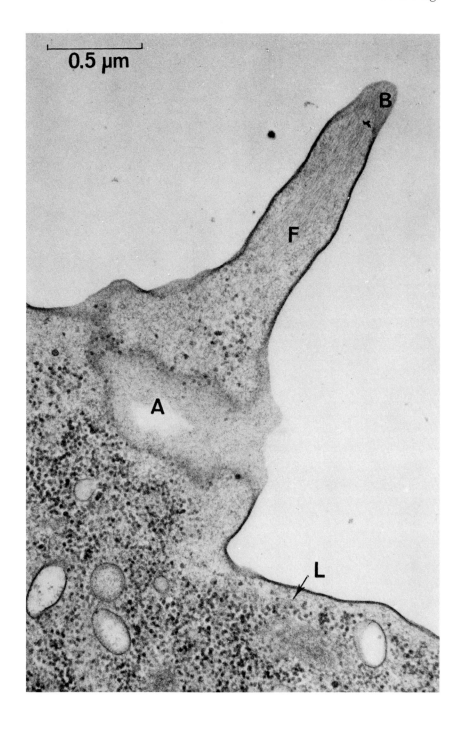

ated muscle (Huxley, 1973). Muscular contraction is thought to result from the relative sliding motion of thick and thin filaments containing myosin and actin, respectively. Evidence for a similar contractile system operating during amoeboid movement in *D. discoideum* rests on the observation that both actin and myosin are found in these motile cells (Woolley, 1972; Clarke and Spudich, 1974). In fact, actin appears to be a major constituent of the soluble proteins of aggregating cells, accounting for about 20% of the newly synthesized proteins at this stage (Tuchman *et al.*, 1974). Some of the actomyosin appears to be bound to the cell membrane (Spudich, 1974). Thin filaments have been seen by electron microscopy that are closely opposed and aligned with the cell membrane in *D. discoideum* (George *et al.*, 1972b; see also Fig. 2.4). This observation suggests that amoeboid movement may result from the relative movement of bound and free actin molecules present in cells with a well-defined polarity (Spudich, 1974). It is clear that until we have some idea of the biochemical basis of amoeboid movement, further analysis of chemotaxis and morphopoeitic movement will be severely limited.

Feeding

Amoebae are attracted to bacteria by a sensory system which appears to involve chemotaxis to folic acid liberated by the bacteria (Bonner *et al.*, 1970; Bonner, 1971). Amoebae ingest their prey by phagocytosis (Fig. 2.5). The bacteria are digested in food vacuoles probably by lytic enzymes liberated into the food vacuoles by fusion of lysosomes. Most of the bacterial material is degraded as very little remains after digestion (Braun *et al.*, 1972). During the process of digestion multilaminar membrane whorls are formed which are ultimately excreted (Fig. 2.6) (Hohl, 1965).

Ultrastructure

A variety of ultrastructural studies of *D. discoideum* have been carried out including those of Gezelius and Rånby (1957), Mercer and Shaffer

---

Fig. 2.4. Aligned microfilaments. A tangential section through a filopodium exposes 44-Å diameter fibers (F) oriented along the long axis and continuous with the peripheral microfibrial layer (L). Regions (A) and (B) are the result of convolutions which extend out of the plane of section. Similar microfilaments can be seen in the peripheral layer of both axenically grown amoebae and amoebae feeding on bacteria (Figs. 2.7 and 2.8) (TEM).

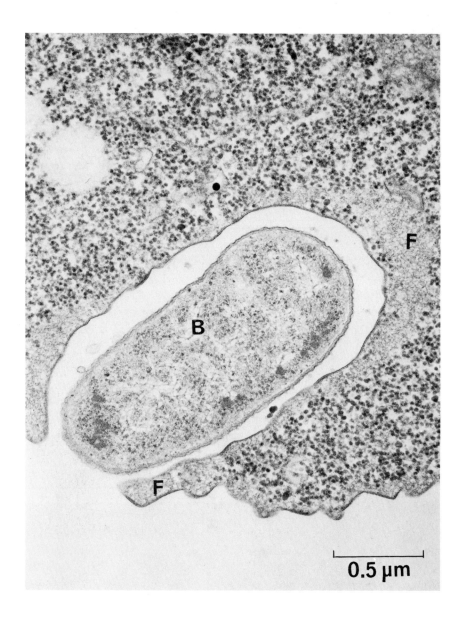

Fig. 2.5. Phagocytosis of a bacterium. An amoeba has almost engulfed a bacterium (B). The peripheral microfibrillar array is extended in localized areas (F) (TEM).

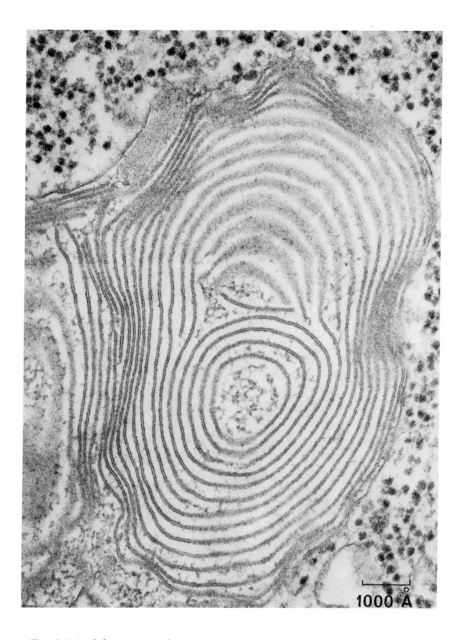

1000 Å

Fig. 2.6. Multilaminar membrane whorls. These myelin figures are found in cells feeding on bacteria but are absent in axenically grown cells (Figs. 2.7 and 2.8). Two 30-Å thick electron dense layers are separated by a 40-Å thick electron transparent layer. This structure differs in ultrastructural dimensions from both mammalian myelin figures and the plasma membrane of *D. discoideum* (Fig. 2.13). It is similar to structures formed by repeated fusion of lysosomes and autophagic vacuoles during spermiogenesis in *Drosophila,* and may arise by a similar mechanism (TEM).

(1960), Maeda and Takeuchi (1969), Hohl and Hamamoto (1969b), Ashworth *et al.* (1969), Gregg and Badman (1970), and George *et al.* (1972b). Each laboratory has used somewhat different fixation and staining procedures. Farnsworth (1973a) has made a systematic analysis of the procedures and has found that best ultrastructural preservation of vegetative amoebae, as well as spores and stalk cells can be had by fixation in 2% glutaraldehyde in 0.05 M phosphate buffer pH 7.4, at 22°C for 60–90 minutes. After washing with 0.1 M phosphate buffer pH 7.4, the material is treated at 4°C with 1% OsO_4 prepared in 0.05 M phosphate buffer pH 7.4 for 45–60 minutes. The fixed material is dehydrated through an ethanol series at 4°C and transferred to absolute ethanol at room temperature. Infiltration of fixed pseudoplasmodia by conventional epoxy resins is adequate only after 16 hours or more at room temperature. Alternatively, the newer low viscosity resins can be used and given adequate infiltration within 8 hours.

Staining of thick sections (0.5 μm thick or greater) (Fig. 7.4) has been found to be satisfactory in about 30 seconds in a 1% solution of toluidine blue in borate at 38°C. Staining of thin sections is optimal after about 2 minutes in saturated uranyl acetate in 50% ethanol, followed by 2 minutes in Reynolds lead citrate strain. Figures 2.7–2.15 were prepared by following these principles.

Fixation and staining of macrocysts (Figs. 4.4 and 4.5) require more rigorous conditions, and it has been found necessary to double the time for primary and postfixations.

Fixation with osmium tetroxide alone (Gezelius and Rånby, 1957; Mercer and Shaffer, 1960; Maeda and Takeuchi, 1969; Maeda, 1971) does not appear to preserve cellular elements as well as glutaraldehyde followed by osmium tetroxide. Fixation with high concentrations of glutaraldehyde (6%) (Hohl and Hamamoto, 1969b) does not appear to improve preservation and leads to cellular damage. Treatment with 2% glutaraldehyde for less than 30 minutes (Gregg and Badman, 1970) is insufficient for complete fixation, while treatment with 2% glutaraldehyde for more than 2 hours can damage vegetative amoebae. Prolonged treatment with osmium tetroxide (Gregg and Badman, 1970) can damage membranes and lead to loss of fine structure.

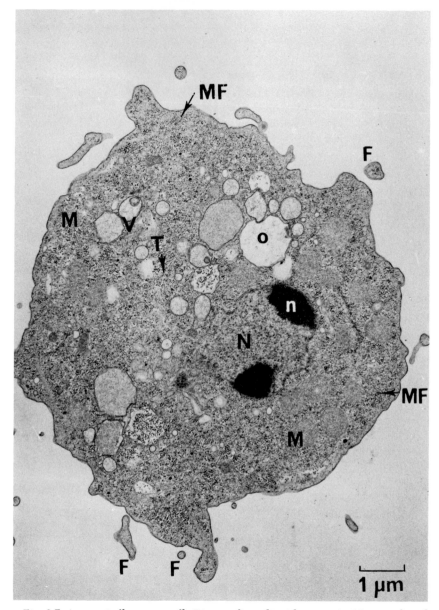

Fig. 2.7. An axenically grown cell. *Dictyostelium discoideum* strain A3 was cultured in suspension in HL-5 broth medium. The cells appear as 6–10 μm diameter spheres with many fine filopodial projections (F). Organelles described in more detail below include: nucleus (N) and nucleolus (n), mitochondria (M), osmoregulatory vacuoles (o), peripheral microfibrillar layer which is 0.3 μm wide (MF), various vesicular structures (V), and microtubular fragment (T) (TEM).

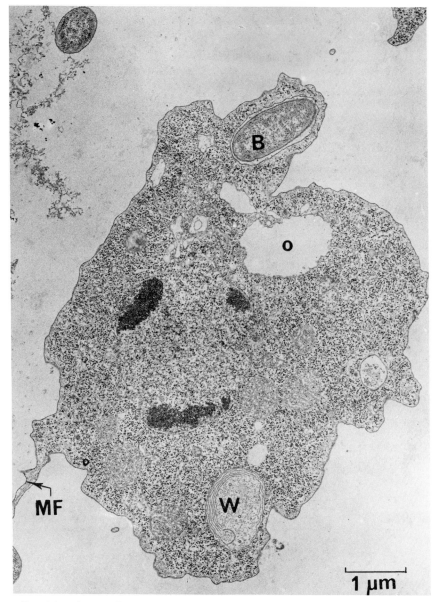

Fig. 2.8. A bacterially grown cell. *Dictylostelium discoideum* strain A3 was grown in association with *K. aerogenes*. A recently engulfed bacterium (B) can be seen (see also Fig. 2.5). A multilaminar whorl (W) is apparent (see also Fig. 2.6). The microfilaments (MF) are visible in the cell projection (see also Fig. 2.4). All the organelles seen in axenically grown cells are also present in these cells; however, the osmoregulatory apparatus (o) is smaller (TEM).

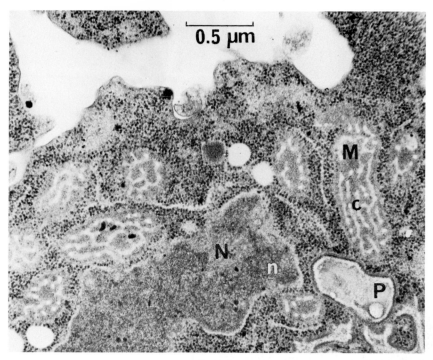

Fig. 2.9. Cellular organelles. Mitochondria (M) appear as cylindrical structures 0.25 μm wide and up to 1.5 μm long containing tubular cristae (c) which are about 300 Å in diameter and are parallel to the long axis of the mitochondrion, often extending the full length and anastamosing. The surrounding membranes consist of two 80-Å wide electron dense layers separated by a 100-Å wide electron transparent region. Intramitochondrial electron dense granules which are 300 Å in diameter can also be seen. About 20 to 30 mitochondrial profiles are found in most medial 1000-Å thick sections. The appearance of the mitochondria does not change during development except that during culmination dense inclusions accumulate and the mitochondria become rounded (see Fig. 7.7B). No evidence for transformation of mitochondria into prespore vesicles has been found although this has been suggested by Maeda (1971). A prespore vesicle (P) can be seen in this cell taken from a pseudoplasmodium. A tangenital section through the nucleus (N) shows the darker polar nucleolar (n) cap almost filling the profile (TEM).

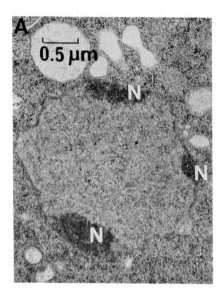
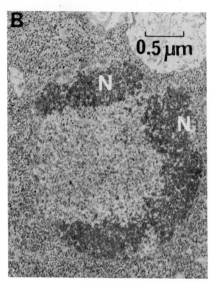

Fig. 2.10. Nuclear region. Nuclei appear as 2–3 μm diameter spheres. Dense nucleolar regions (N) at the periphery of nuclei are seen in all sections, indicating that much of the immediate inner surface is covered by this organelle. The regions are often distinct (Fig. 2.10A) but are also seen to be contiguous (Fig. 2.10B) which suggests that the nucleolus is a multilobose structure. In some regions the nucleoli appear fibrous while in others they appear granular. No significant change in ultrastructure of the nucleoli has been observed during development, in contrast to the report of Madea and Takeuchi (1969) that the nucleoli become more granular as development proceeds (TEM).

Fig. 2.11. Endoplasmic reticulum. The electron dense spheres, 180–200 Å in diameter, are ribosomes. They are frequently attached to long straight regions of endoplasmic reticulum (arrows). (See also Fig. 7.7.) Such straight endoplasmic reticulum is an unusual feature in cells of most species but is common in *D. discoideum* (TEM).

Fig. 2.12. Plasma membrane. A very thin section (∼ 400 Å) stained with uranyl acetate/lead citrate shows the marked asymmetry of the plasma membrane. The outer electron dense layer is 40 Å wide while the inner electron dense layer is 26 Å wide. These layers are separated by an electron transparent layer 30 Å wide. The variations along the membrane are due to the angle of sectioning (TEM).

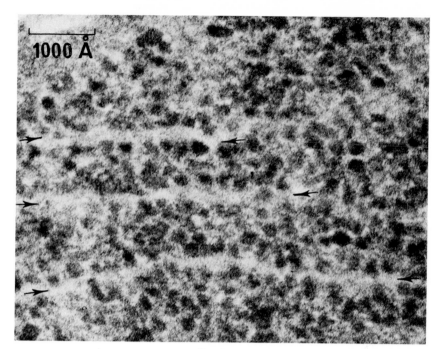

Fig. 2.11

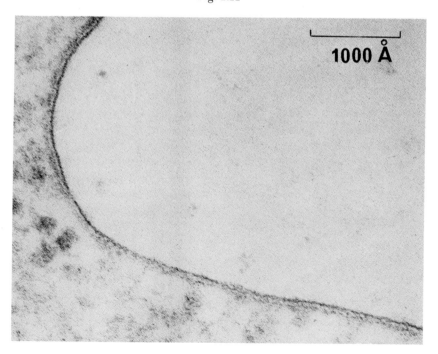

Fig. 2.12

Fig. 2.13. Crystal bodies. These structures, which appear to be cuboidal crystals of a 0.5-μm side, are found in cells from all developmental stages including spores and stalk cells. The crystals display an asymmetric 90 by 105 Å lattice, and may be composed of protein. Similar structures with identical spacing have been seen in the liver of the slender salamander *Batrachoseps attenuatis* and in the interstitial cells of the human testes (Fawcett, 1966) (TEM).

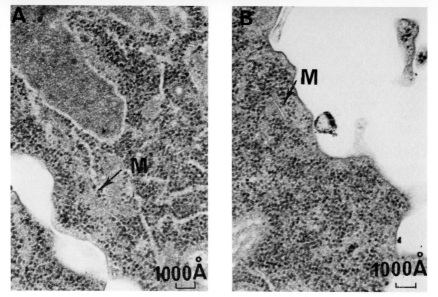

Fig. 2.14. Microtubules. Microtubular fragments (M) can be seen in two sections of cells taken from migrating pseudoplasmodia. They are 120 Å in diameter and are seldom found to be longer than the fragment seen in A. They cannot be observed in material fixed at 4°C (TEM).

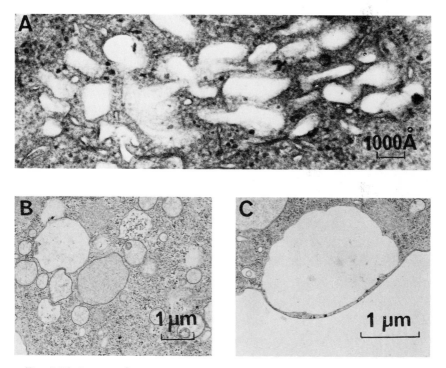

Fig. 2.15. Osmoregulatory apparatus. (A) Many small vesicles about 0.25 μm in diameter appear to fuse to form small vacuoles; (B) these vacuoles fuse and become larger; (C) finally a vacuole 2–3 μm in diameter is formed before egestion (TEM).

Chapter 3

The Genome

Karyotype

The original isolate of *D. discoideum* NC-4 has been reported to contain 7 chromosomes (Wilson and Ross, 1957; Sussman, 1961a,b). Since the organism appears to carry an uneven number of chromosomes it is likely that it occurs in a haploid state. The chromosomes are difficult to visualize but can be stained by Giemsa after the cells are fixed in methanol-acetic acid. Diploid strains containing 14 chromosomes have been isolated from *D. discoideum* NC-4 and have been found to give rise to haploid progeny at variable rates (R. Sussman and Sussman, 1963). Some diploid strains segregate haploids only rarely while others give rise to haploids at a rate such that they make up to 50% of the population after only 40 generations (Loomis, 1969b). The mechanism which generates diploids in *D. discoideum* NC-4 is poorly understood but appears to involve anastomosis of cells followed by karyogamy (Huffman *et al.*, 1962). Segregation of haploid from diploid strains appears to occur by a mechanism similar to the process of chromosomal reduction which occurs in interspecies mammalian hybrids formed by cell fusion. Individual chromosomes are lost at cell division, giving rise to aneuploid progeny which undergo further chromosome loss during subsequent divisions until the haploid complement is reached. The diploid strains are further characterized by being more resistant than the haploid strain to killing by ultraviolet light (Freim and Deering, 1970: Jobe and Loomis, unpublished). Moreover, the spores formed by diploid strains are considerably larger than those formed by haploid strains (Sussman and Sussman, 1963; Sackin and Ashworth, 1969).

Nuclear Composition

Nuclei of *D. discoideum* contain DNA, RNA, and protein in a ratio of 1:3.4:22 (Coukell and Walker, 1973). The basic nuclear proteins have been analyzed electrophoretically on acrylamide gels and found to have properties similar to those of histones found in calf thymus nuclei (Coukell and Walker, 1973). Five major bands were observed on SDS-acrylamide gels of which 2 co-migrated with thymus histones f2a1 and f3. Moreover, *D. discoideum* histones which migrated in the region of lysine-rich calf thymus histone f1 could be resolved into several subfractions on urea gels. Differences in the complement of f1 histones have been found in tissues of different origin as well as in the same tissue in different metabolic states, and it has been suggested that these histones may play a role in differential genetic expression.

Axenically grown cells of *D. discoideum* contain about 10^{-13} gm DNA per cell (Ashworth and Watts, 1970; R. R. Sussman and Rayner, 1971; Leach and Ashworth, 1972). Because of the low DNA content of the organism and the small number of chromosomes, *Dictyostelium* provides excellent material for analysis of the composition and topology of a eukaryotic genome. The nuclear genome size is estimated from chemical analysis to be about 3.6×10^{10} daltons or about 5.4×10^{7} nucleotide pairs (R. R. Sussman and Rayner, 1971). Whole nuclear DNA has a buoyant density of 1.683 gm cm^{-3} and a T_m of 78.8°C indicating a base composition of 23% G + C (guanine plus cytosine) (R. R. Sussman and Rayner, 1971; Firtel and Bonner, 1972a; Leach and Ashworth, 1972). The nuclear DNA can be separated by CsCl density-gradient centrifugation into a main band and a pair of satellites (Firtel and Bonner, 1972a,b).

Mitochondrial DNA

DNA has been isolated from purified mitochondria and found to make up 28–40% of the total cellular DNA (R. R. Sussman and Rayner, 1971;

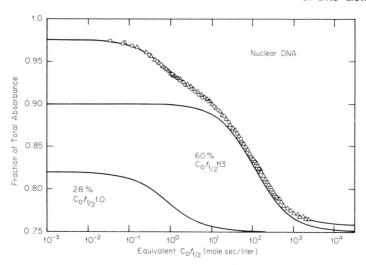

Fig. 3.1. Reassociation kinetics of nuclear DNA. DNA was extracted from nuclei and purified by isopycnic centrifugation. After shearing, the DNA was dissociated by heating and allowed to reanneal. A solution of 105 μg/ml DNA was incubated at 56°C in 0.12 M phosphate buffer (○); another solution containing 1150 μg/ml DNA was incubated at 60°C in 0.24 M phosphate buffer (△). Reassociation was measured optically. The data was subjected to computer analysis to yield the kinetics of reannealing of two separate fractions (Firtel and Bonner, 1972a).

Firtel and Bonner, 1972a). This DNA has a T_m of 80°C and a buoyant density indicating a base composition of 28% G + C. No evidence for circularity of mitochondrial DNA has been found in this organism; however, growth in the presence of ethidium bromide results in cells depleted of mitochondrial DNA (Firtel and Bonner, 1972a). Ethidium bromide is known to intercalate preferentially into twisted circular DNA and to inhibit replication. Thus, the preferential effect of this drug on mitochondrial DNA suggests a structure different from that of nuclear DNA.

The kinetics of reannealing of thermally dissociated DNA strands has given evidence on the complexity of both nuclear and mitochondrial DNA. When DNA is sheared to segments of about 450 bases and the strands are separated, complementary sequences will reanneal to form double strands under appropriate conditions dependent on both the initial concentration of the fragments (C_0) and the time (t) of reaction. Repetitive segments present in several copies in a cell will reanneal more rapidly than unique sequences present only once per genome. A $C_0 t$ analysis of mitochondrial DNA has indicated that each cell contains about 200–250 identical copies of segments of DNA of about 35–40×10^6 daltons (Firtel and Bonner, 1972a). This DNA is sufficient to code for about 50 different RNA molecules of 1000 nucleotides each.

Repetitive Sequences

Nuclear DNA reanneals in two distinct stages (Fig. 3.1) (Firtel and Bonner, 1972a). A rapidly annealing fraction makes up about a third of the genome and is characterized by a $C_0t_{1/2}$ equal to 1.0 (the C_0t value for half reannealing). Some of the purified reannealed repetitive DNA can be melted at lower temperatures than unique sequences but the overall T_m is similar in both fractions. This suggests that the base mismatching in reannealed repetitive DNA is not dramatically different from that in reannealed unique DNA. The function of this repetitive fraction of the genome is presently a matter of active concern. It appears to be made up of families of sequences present at different frequencies. Some of these sequences are interspersed throughout the chromosomes (Firtel *et al.*, 1975) and may carry out regulatory functions. Other sequences may serve control functions of some unknown sort, while still others may be necessary for chromosomal functions such as replication and mitotic segregation.

The genes which code for ribosomal RNA (rRNA) make up a small but well-characterized proportion of the repetitive DNA. The sequences which code for the 27 S and 17 S rRNA make up 2.2% of the nuclear DNA. The rDNA sequences have a buoyant density in CsCl of approximately 1.682 gm cm⁻³ and band in CsCl density gradients at the same position as mitochondrial DNA (Firtel and Bonner, 1972a). rDNA has an apparent $G + C$ content of 28%, similar to that of whole cell DNA. Since the $G + C$ content of the products of rDNA and 27 S and 17 S rRNA is 42% (Pannbacker, 1966), to account for the buoyant density these products must be coded for by transcribed regions attached to spacer regions which are $A + T$ rich. The ribosomal sequences can be partially separated from bulk DNA either by hybridization to purified rRNA or by centrifugation in $AgSO_4/CsSO_4$ density equilibrium gradients (Firtel and Bonner, 1972a; Miller and Loomis, unpublished). There appear to be about 150 copies of rDNA per genome.

Unique Sequences

When the slowly reannealing fraction of nuclear DNA is purified, it reassociates with a $C_0t_{1/2}$ equal to about 70. From the size of the genome, it can be estimated that sequences annealing at this rate are present only once in the genome. These single copy sequences make up about two-thirds of the nuclear DNA. The complexity of the single-copy sequences is about 7 times that of *E. coli* or approximately 20×10^9 daltons (Firtel and Bonner, 1972a). This is sufficient to code for about 28,000 different RNA molecules of 1000 nucleotides each. All but a few percent of the mRNA found in either growing or developing cells hybridizes exclusively with the single-copy fraction of nuclear DNA (Firtel *et al.*, 1972; Firtel and Lodish, 1973). Thus, it is clear that the unique sequences of the genome of *D. discoideum* code for the gene products synthesized during both growth and development. This characteristic has been utilized to characterize the portions of the genome expressed at different stages, as is discussed in Chapter 8.

Chapter 4

Genetics of *Dictyostelium*

Genetic Approaches

Genetic analyses have been essential for the elucidation of the metabolic pathways and control mechanisms of bacteria and viruses (Jacob
and Monod, 1961; Epstein *et al.*, 1963). The application of similar approaches to developing systems holds much promise for clarifying the
essential steps in programmed synthesis and morphogenesis. *Dictyostelium* presents a highly favorable system in which to develop such an
approach due to the fact that it is haploid and has a unique life cycle
which includes a single-celled growth phase and a multicellular differentiation phase.

Genetic analyses of *D. discoideum* have followed several approaches
which have shed light on developmental processes. One approach has
been to isolate a large number of morphological mutants so as to give
some indication of the number of variations open to the developing
system and to determine causal connections between stages. For example,
it has been consistently observed that mutant strains unable to aggregate

fail to form either spore cells or stalk cells. Clearly, some process dependent on the formation of multicellular aggregates is required for the terminal differentiation of the cells. Another approach has been to isolate mutations in specific gene products known to accumulate during unique stages. The effect of these mutations on other molecular differentiations as well as on the overall developmental plan has indicated the physiological function of the specific gene products.

The formation of diploid strains from pairs of dissimilar haploid strains and subsequent analysis of expression of the original markers have been shown to be possible. This technique can be used to indicate the dominance and *cis* effects of specific mutations. Moreover, genetic exchange and segregation of markers can separate mutations and place them in homologous genetic backgrounds where their pleiotropic effects can be studied unhindered by the effects of secondary unselected mutations. Segregation of haploid progeny from heterozygous strains has also been used to construct genetic linkage maps.

Mutational Genetics

A large series of morphological mutants of *Dictyostelium* has been isolated in which almost all of the gross aspects of morphogenesis are affected in one strain or another (Table 4.1). The detailed analysis of the mutants are discussed in subsequent sections concerning the affected stages.

The utilization of haploid strains and powerful mutagens such as N-methyl N'-nitronitrosoguanidine or ethyl methane sulfonate allows one to recover a high proportion of morphological mutants in a mutagenized population (Yanagisawa *et al.*, 1967). When less than a hundred spores or amoebae are spread on a plate with the bacterial associate, the amoebae grow as discrete clones, clearing a circular plaque in the bacterial lawn (Fig. 4.1). This growth pattern allows one to isolate genetically different strains present in a population. Mutations affecting multicellular developmental processes can be selected by this technique since cells of each clone will aggregate and form genetically homogeneous pseudoplasmodia. Genetic defects in morphogenesis can be recognized by visually screening the developmental stages in the individual plaques. Using this technique a large number of morphological mutants have been isolated (Sussman, 1955a; Yanagisawa *et al.*, 1967). Mutations affecting plaque formation can also be screened by this technique, and this approach has turned up a series of minute plaque-forming strains which grow much more slowly than the wild-type strain and so form only pinpricks in the bacterial lawn after 3 days of growth (Fig. 4.1)

TABLE 4.1
MORPHOLOGICAL MUTANTS

State affected	Designation[a]	Phenotypic characteristics[b]
Aggregation	Aggregateless	Failure to produce cAMP (1,1b)
	Aggregateless	Failure to respond to cAMP (1b)
	Aggregateless	Failure to either produce or respond to cAMP (1b)
	Aggregateless (Agg50)	Overproduction of cAMP phosphodiesterase (2)
	Giant aggregation territories (ga86)	Low phosphodiesterase (2)
	Aggregateless (aggr50-2)	Failure to produce inhibitor of phosphodiesterase (3)
	Aggregateless (wag 2)	Failure to produce cell-bound phosphodiesterase (4)
	Aggregateless (WL-5)	Failure to produce cohesion protein (5)
	Fruity/dwarf (Fty-1)	Small aggregation territories (1,6)
Pseudoplasmodium	Bushy (bu)	Increased number of papillae (6)
	Fruitless (Fts-1)	No conversion to pseudoplasmodia (1)
	Blind (L-25)	Failure in phototaxis (7)
	Crippled (DBL211)	Failure in migration (8)
Culmination	Stalkless (KY-19)	Reduced stalk formation (9)
	Sporeless (min 2)	Failure to form resistant spores (10)
	Fast (FR-17)	Disorganized fruiting bodies; rapid development (11)
	Slow (GN-3)	Delayed morphogenesis (12)
	Hairy (KY-3)	Stalked migration; no culmination (9)
	Hedgehog	Multiple stalks arising from a pseudoplasmodium (13)
	Slippery stalks	Sorus found at base of stalk (13)
	White (I-262)	Failure to produce carotenoid (14)
	Brown (I-47)	Secretion of brown pigment (14)

[a] When a specific mutant strain of the class has been well characterized, it is given in parentheses.

[b] Key to references:

(1) Sussman and Sussman (1953)
(1b) Bonner *et al.* (1969)
(2) Riedel and Gerisch (1971)
(3) Riedel *et al.* (1973)
(4) Malchow *et al.* (1972)
(5) Rosen and Loomis (unpublished)
(6) Sussman (1955a)
(7) Loomis (1970b)

(8) Dimond *et al.* (1973)
(9) Yanagisawa *et al.* (1967)
(10) Loomis (1968)
(11) Sonneborn *et al.* (1963)
(12) Loomis (1970c)
(13) Loomis, (unpublished)
(14) Sussman and Sussman (1963)

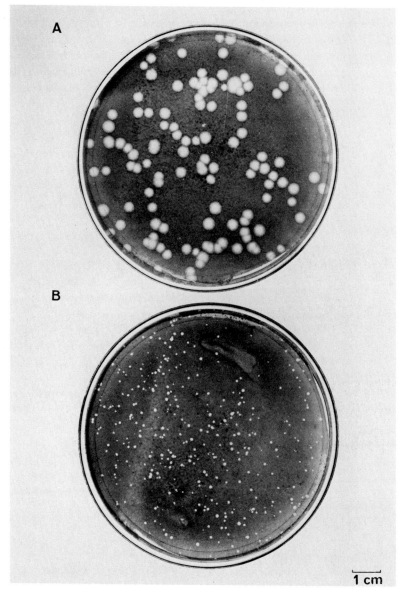

Fig. 4.1. Plaques in a bacterial lawn. The clear areas are regions in which most of the bacteria have been ingested by cells of *D. discoideum* growing clonally on 9 cm diameter petri dishes spread with 10^8 *K. aerogenes*. (A) Plaques formed by the wild-type strain NC-4 after incubation at 22°C for 3 days; (B) plaques formed by the *minute* mutant strain *min* 2 after incubation at 22°C for 4 days.

(Loomis and Ashworth, 1968). Temperature-sensitive mutants have also been recovered which do not grow at all at 27°C but grow well at 22°C (Loomis, 1969b; Katz and Sussman, 1972). These strains have allowed studies on genetic exchange and recombination which will be discussed below.

Clonal growth of large numbers of axenic strains can be conveniently produced by inoculating 0.1 ml of medium containing a single spore or cell into the wells of plastic multitest plates. Maximal growth and the formation of visible colonies occur in about 2–3 weeks. Since hundreds of plates, each containing 96 wells, can be easily inoculated, the technique has permitted the analysis of tens of thousands of clones (Dimond *et al.*, 1973).

When a population is treated with nitrosoguanidine such that only one cell in a thousand survives, one can calculate from a Poisson distribution that there is an average of 7 lethal mutations per cell (Dimond *et al.*, 1973). In such a population of mutagenized cells, about 30% of the surviving cells exhibit a visible morphological aberration. This suggests that only about a twentieth as many genes are involved in the gross aspects of morphogenesis as are necessary for growth. As will be discussed in Chapter 8, the estimated number of genes expressed uniquely during development is similar to that expressed during growth, about 8000 genes. This discrepancy suggests that the majority of genes expressed during development play a subtle supportive role rather than functioning as required steps in a strictly linear pathway to the terminal differentiations.

Parasexuality

The life cycle of *D. discoideum* contains no obligatory sexual phase. Almost all of the haploid cells enter into aggregates and retain their individuality throughout morphogenesis and give rise to haploid spores. On occasion cells have been observed to engulf one another in the population (Huffman *et al.*, 1962). The fate of the assimilated nucleus could not be determined by the cytological methods employed. It could either be destroyed or retained in a heterokaryotic cell or undergo karyogamy to give rise to a diploid cell. The answer has come from the use of genetically marked strains which have shown that pairs of haploid cells form true diploids at a frequency of about 1 in 10,000 cells. R. Sussman and Sussman (1963) described a diploid strain which arose from a mixed population consisting of an albino strain and a brown strain. The diploid was characterized by large spore size, lack of excretion of the brown pigment, and a yellow sorus. This strain segregated both albino and brown haploid progeny at a low frequency. It was suggested that the

TABLE 4.2
Proposed Genetic Nomenclature[a]

Gene symbol[b]	Mutant phenotype	Wild-type phenotype
min	Minute plaque	Large plaque
tsg	Temperature-sensitive growth (27°C)	Growth at 27°C
axe	Axenic growth capacity	Failure to grow axenically
cyc	Cycloheximide resistance (500 μg/ml)	Cycloheximide sensitive (500 mg/ml)
acr	Acridine resistance (100 μg/ml)	Acridine sensitive (100 μg/ml)
fur	5-Fluorouracil resistance (200 μg/ml)	5-Fluorouracil sensitive (200 μg/ml)
fud	5-Fluorodeoxyuridine resistant (200 μg/ml)	5-Fluorodeoxyuridine sensitve (200 μg/ml)
bud	5-Bromodeoxyuridine resistance (2 mg/ml)	5-Bromodeoxyuridine sensitive (2 mg/ml)
spr	Round spores	Ellipsoid spores
whi	White sori	Yellow sori
brw	Brown pigment excretion	No pigment excretion
agg	Aggregateless	Aggregates
frt	Aberrant fruiting body	Normal fruiting body
nag	Affecting N-acetylglucosaminidase	Normal N-acetylglucosaminidase
man	Affecting α-mannosidase	Normal α-mannosidase
glu	Affecting β-glucosidase	Normal β-glucosidase
alp	Affecting alkaline phosphatase	Normal alkaline phosphatase
upp	Affecting UDPG pyrophosphorylase	Normal UDPG pyrophosphorylase

[a] The symbols have been chosen to conform with the system proposed by Demerec *et al.* (1966), and follow the suggestions of Kessin *et al.*, (1974).

[b] Mutations defining these genes have been isolated and are described in Chapters 2, 4, and 9. The wild-type allele should be designated by a superscript +, e.g., *min*+. Several genes have been assigned to linkage groups as discussed below. In many instances more than one complementation group has been defined and these are distinguished by a capital letter following the three letter symbol, e.g., *tsg*A and *tsg*B. The genes coding for specific enzymes will undoubtedly include more than one complementation group since distinct isozymes have been recognized. These should also be designated by capital letters, e.g., *man*A and *man*B.

albino stain carried a recessive mutation *whi* and the brown stain carried a recessive mutation *bwn*. The genetic symbols for these and other genes are given in Table 4.2. The observed phenotypes of the diploid and subsequently segregated haploid strains suggested that the cross could be described as

Parental strains:	*whi*+; *bwn* × *whi*; *bwn*+
Diploid:	[*whi*+/*whi*; *bwn*/*bwn*+]
Haploid segregants:	*whi*+; *bwn* *whi*; *bwn*+ *whi*+; *bwn*+ *whi*; *bwn*

As indicated, the segregants also included yellow-brown and albino brown strains at a high frequency suggesting that the genetic markers had segregated independently.

These observations indicated that genetic exchange can occur in *D. discoideum* and opened up the possibility of segregational analyses but were unable to define the mechanism or frequency at which the events occurred. To reproducibly observe the formation of heterozygous diploids, a selection procedure was necessary to collect the rare diploids. Two procedures have been developed.

The first system which could reproducibly analyze genetic exchange utilized *min* strains which form minute plaques (Loomis and Ashworth, 1969). Plates could be spread with spores from a fruiting body formed from cells of two independently isolated minute (*min*) strains. After 3 days several thousand minute plaques would form in the bacterial lawn. At a frequency of about 10^{-4}, large plaques were found to arise which, by spore size, were found to be diploid. The heterozygous diploid strains bred true for many generations. When the *min* strains carried other genetic markers such as resistance to cycloheximide, *cyc*, or a defect in aggregation, *agg*, the diploid was found to extinguish these characteristics, indicating they arose from recessive mutations. A typical cross and subsequently observed segregation is

Parental
   strains: ($minA$ $minB^+$ cyc^+ agg^+) × ($minA^+$ $minB$ cyc agg)
Diploid: [$minA/minA^+$; $minB^+/minB$; cyc^+/cyc; agg^+/agg]
Haploid
   segregants: ($minA$ $minB^+$ cyc agg^+); ($minA^+$ $minB^+$ cyc^+ agg^+); ($minA^+$ $minB^+$ cyc^+ agg)

The original large-plaque-forming diploid stain was wild type with respect to both aggregation and drug sensitivity, and the majority of the subclones had similar phenotypes. However, a few clones had recombinant phenotypes in which various parental markers were expressed. The analysis of this pattern indicated independent segregation of the genes for plaque size, cycloheximide resistance, and morphogenesis. It is likely that the original diploid gave rise to haploid or aneuploid progeny at a low frequency in which individual chromosomes were randomly reassorted. This parasexual process is indicated in Fig. 4.2.

Sinha and Ashworth (1969) provided further evidence for the parasexual cycle in studies involving several other markers in the *min* strains as well as karyotypic analysis of the recombinants. The original *min* strains were found to contain 7 chromosomes while the heterozygous diploid strains were uninucleated and contained 14 chromosomes. Segre-

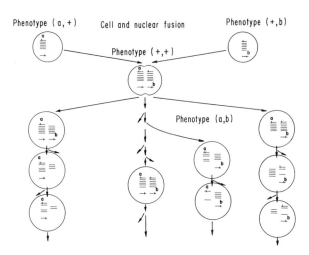

Fig. 4.2. Genetic exchange in *D. discoideum*. Haploid strains carrying independent genetic markers, a and b, form diploids at a low frequency. Lines represent individual chromosomes. The diploids propagate but also lose chromosomes at a low frequency giving rise to recombinant aneuploid and haploid progeny.

gants contained from 7 to 13 chromosomes. It was concluded that chromosomes are lost randomly from *D. discoideum* diploids during subculture as seems to occur in interspecies somatic crosses. The system has been compared to the parasexual cycle of *Aspergillus* (Sinha and Ashworth, 1969).

Although heterozygous diploids can be routinely recovered from pairs of *min* strains, it has been difficult to quantitate the frequency of genetic exchange and subsequent segregation using this system because of the large differences in growth rate of the mutant and wild-type isolates. For this reason conditional mutants were sought from which one could recover heterozygous strains. Several temperature-sensitive strains were isolated which grow at 22°C, but not at 27°C (Loomis, 1969b). The growth of an axenic derivative of one of these strains is shown in Fig. 4.3. Different mutant strains carry independent *tsg* mutations affecting vital functions and so will complement in a heterozygous diploid. By selecting for strains able to grow at 27°C which arose from a pair of strains carrying *tsg* mutations, it was possible to delineate the time and frequency of genetic exchange and make an estimate of the segregation frequency. A period of close cell association such as that provided by aggregation was found to be required for genetic exchange, but neither migration nor fruiting body formation was essential. Sixteen hours after the initiation of development the frequency of temperature-resistant

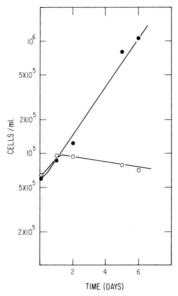

Fig. 4.3. Growth of strain GTS7a. Amoebae of the temperature-sensitive strain GTS7a were inoculated into HL-5 medium and incubated either at 22°C (●——●) or 27°C (○——○). Cell counts were made on a Coulter Counter. Strain GTS7a is an axenic derivative of strain GTS7 (Loomis, 1969b).

(TR) strains in a population of 2 *tsg* strains was 6×10^{-5} (Loomis, 1969b). Moreover, 2 aggregateless *tsg* strains gave rise to temperature-resistant progeny at a frequency of 3×10^{-4} when allowed to lie together in amorphous mounds for 24 hours. These TR strains, unlike the parental strains, developed normally. It was concluded that the original *agg* mutations were recessive.

The rate of segregation of aneuploid and haploid strains has been difficult to estimate in any of these systems. R. Sussman and Sussman (1963) found that segregants made up 18% of the population after 23 generations of growth of their most stable diploid, H-1. Most of the heterozygous strains from the *agg* × *tsg* crosses mentioned above contained about 30% segregants after 30 generations (Loomis, 1969b). However, Katz and Sussman (1972), using *tsg* mutants isolated from the spore-color strains previously used by R. Sussman and Sussman (1963) and carrying a recessive cycloheximide resistant marker, found segregation of drug-resistant strains at a frequency of only 0.03%.

As R. Sussman and Sussman (1963) have shown, haploid segregants are often favored by a growth advantage over diploid strains. Within as little as 20 generations, haploid segregants initially present at less than

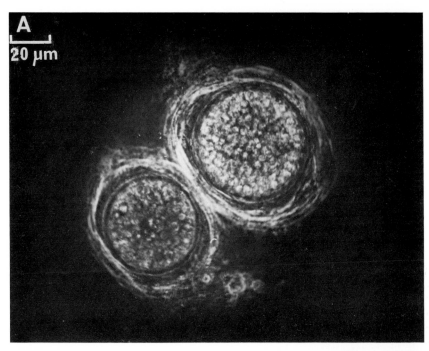

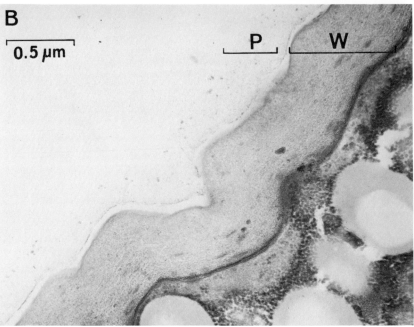

1% can overtake the culture and make up to 20% of the population. Thus, it is difficult to know the true segregation frequency. Likewise, the low figure found by Katz and Sussman (1972) may not give a true estimation of the segregation frequency since it is equally possible that the parental strains used in this cross carried mutations resulting in impaired growth of the haploids at the permissive temperature. Segregants would then be diluted out of the population by the "hybrid vigor" of the heterozygous diploids. Clearly single generation segregation analysis must be done before the actual segregation rate can be estimated.

Segregational analysis of the cross of *tsg* derivatives of the spore-color mutants has indicated that the white allele (*whi*) is linked to one of the *tsg* alleles (Katz and Sussman, 1972). These markers segregate together but independently of other markers and thus comprise the first linkage group. Recent studies have extended this linkage group which now includes other temperature-sensitive markers, a gene necessary for growth in broth medium (*axe*A), as well as a gene conferring resistance to both methanol and acridine (*acr*A) (Gingold and Ashworth, 1974; Williams *et al.*, 1974). Another linkage group carries genes affecting spore shape (*spr*), resistance to acridine (*acr*B), resistance to cycloheximide (*cyc*A), and temperature sensitivity (*tsg*E).

There are now five linkage groups (Kessin *et al.*, 1974):

I:	*spr*	*cyc*A	*tsg*E	*acr*B
II:	*whi*	*axe*A	*tsg*D	*acr*A
III:	*tsg*A	*axe*B		
IV:	*bwn*	*tsg*B		
V:	*tsg*C			

Gingold and Ashworth (1974) have been able to show that mitotic crossing over occurs in heterozygous diploid strains of *D. discoideum*

Fig. 4.4 Macrocysts of *D. discoideum*. (A) Spherical clumps are formed when cells of different mating types are mixed. The cysts are up to 100 μm in diameter. Several peripheral layers surround the cysts (phase contrast optics). (B) The peripheral layers can be seen to be composed of a loose, fibrous outer primary wall (P) which can be up to 20 μm thick and a denser inner wall (W) which is about 0.5 μm thick. The primary wall is composed of material similar in appearance to the surface sheath which surrounds pseudoplasmodia (see Fig. 6.4). The inner wall consists of an outer electron-dense 200 Å thick layer overlying the 0.5 μm thick fibrous region which is presumed to contain a high content of cellulose (Blaskovics and Raper, 1957). There is an inner electron-dense layer 250 Å thick which is closely opposed to the limiting membrane of the macrocyst. The total wall structure is similar to that of mature spores (see Fig. 7.7). Erdos *et al.* (1973a) refer to the inner layer of the cyst wall as a separate "tertiary wall," but it appears that the separation in their preparation may be a fixation artifact and that this layer is not a cytologically distinct structure (TEM).

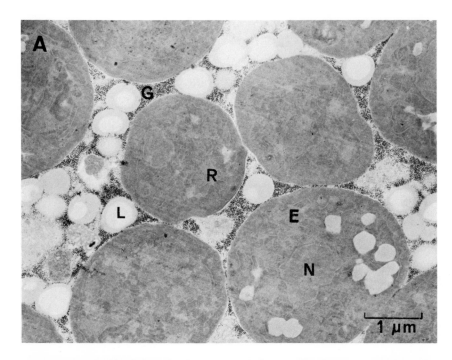

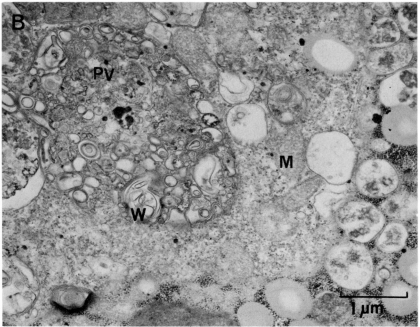

which results in progeny which are partially homozygous. They were able to separate markers on two linkage groups but were unable to definitively order the markers because the frequencies at which various combinations were recovered were affected by the presence of mutations which altered the growth rate of the segregants. The markers will have to be placed in a common genetic background before definitive mapping can be done by the parasexual system.

Further segregational analysis using a variety of markers promises to extend the map so that each chromosome can be conveniently manipulated. This will be important when the dominance or *cis* effect of a given mutation is to be investigated. Possible markers include auxotrophic ones as well as drug-resistant markers. Axenic strains have been isolated which are resistant either to BUdR (bromodeoxyuridine), FUdR (fluorodeoxyuridine), or FU (fluorouracil) (Loomis, 1971).

Sexuality

Although there is no evidence for sexual mating types in strain NC-4, when this strain is mixed with several strains isolated from other geographical sites, a mating type reaction results in the formation of macrocysts (Erdos *et al.*, 1973b; Clark *et al.*, 1973). When cells of strain NC-4 and strain V-12 are mixed and incubated in submerged culture in the absence of nutrients, the cells aggregate into roughly spherical clumps within which macrocysts develop (Fig. 4.4). These consist of groups of about 100 cells surrounded by a double wall. It has been reported that a giant cell develops in the center of the cyst which engulfs the remaining cells before going through a dormant stage (Erdos *et al.*, 1973a,b). After a period of a week or more, the giant cell divides to produce a population of endocytes (Fig. 4.5). To date there is only circumstantial evidence that mating has occurred within the macrocysts and that the endocytes are the products of a meiotic division followed by mitotic divisions (Erdos *et al.*, 1973a,b). However, the analysis is now being extended to appropriately marked strains which should indicate whether recombination occurs in the macrocysts.

Fig. 4.5. Fine structure of a mature cyst. (A) Endocytes (E) are embedded in a matrix containing lipid droplets (L) and glycogen granules (G). Preservation of structure is difficult in cysts; however, nuclei (N) and endoplasmic reticulum (ER) can be distinguished. (B) Extensive autodegradation occurs in developing cysts as indicated by the presence of polyvesicular bodies (PV). Degenerating mitochondria (M) and a membrane whorl (W) can be seen (TEM).

Meiotic recombination in *D. discoideum* will greatly facilitate the study of various mutations in diverse genetic backgrounds and will allow detailed genetic mapping. However, dominance relationships and *cis* effects cannot be studied in the mating type system because the apparent diploid stage is restricted to the giant cell of the macrocyst. On the other hand, the parasexual system allows the convenient selection of heterozygous diploids able to carry out all the processes of the developmental cycle.

Stock Maintenance

Analysis of physiological processes by specific mutant strains and the construction of a detailed genetic map require the maintenance of hundreds of well-characterized strains. These can be passed from one growth plate to another indefinitely, but this is tedious and must be done every few weeks. The spores of developmentally competent strains, however, can be freeze-dried and stored in vacuum for up to 10 years with excellent recovery. The spores can also be stored dry on silica gel and are often viable for several years in this form. To resuscitate the strains only a few spore-encrusted silica grains need be spread on a growth plate with a bacterial associate for growth to occur. Some mutant strains which fail to form spores when incubated alone can be induced to form spores when mutant and wild-type cells are mixed and allowed to develop together (see Chapter 12). The spores can then be stored conveniently either on silica gel or in vacuum. The mutant strain can be recovered by spreading only a few of the stored spores on a growth plate and screening the resultant clones for the mutant and wild-type phenotypes. A third way to store stocks is in 10% dimethyl sulfoxide at −70°C. This technique has successfully stored amoebae of aggregateless strains for up to a year and is thus useful for strains unable to form spores.

Chapter 5

Aggregation Stage

Aggregation of tens of thousands of individual cells to a common center is a fascinating problem in cell communication found in few organisms other than the Acrasiales. When placed on a moist surface free of nutrients, the cells of *D. discoideum* initially move about at random. However, after about 6–10 hours, centers are formed to which surrounding cells are attracted. The cells stream into the centers in large wheellike patterns up to several centimeters across for a period of up to 10 hours depending on the cell density (Fig. 5.1). When the area is depleted of cells, the streams move into the center forming a mass about 0.5 mm across.

Chemotaxis

More than 25 years ago, John Bonner demonstrated in an elegant experiment that the aggregation of *D. discoideum* involved chemotaxis to a compound excreted by the cells, which he termed "acrasin" (Bonner,

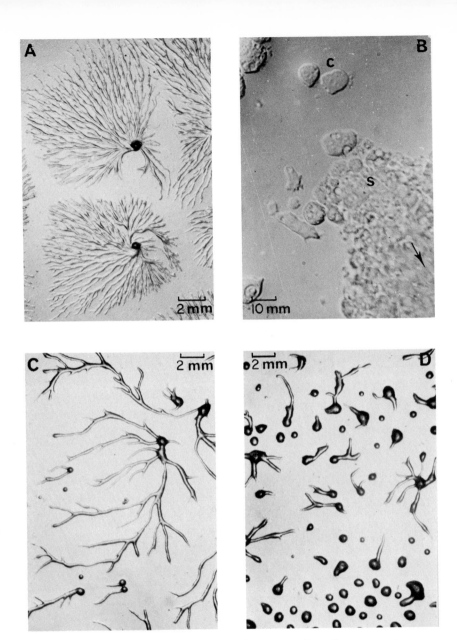

Fig. 5.1. Aggregation. (A) Cells respond chemotactically and aggregate in streams to centers. (B) Individual cells (c) adhere to the streams (s) and proceed toward the center (arrow). When the cells are undisturbed by a cover glass and observed by Nomarski optics, no asymmetry can be seen in the cells before they attach to the streams. Previous reports of bipolarity in the cells about to enter streams have been based on observation of cells under cover glasses (Beug *et al.*, 1970). (C) As aggregation proceeds, the minor branches flow into the major streams. (D) Toward the end of aggregation, cells in the streams all reach the center where they become integrated into pseudoplasmodia (see Figs. 1.6 and 1.7).

1947, 1949). In the experiment, a group of cells was surrounded by responsive cells and positioned in a slowly flowing stream of water. Cells downstream aggregated to the clump of cells while those upstream seemed to be unaware of the presence of the clump. Clearly the central group of cells was excreting a soluble substance which was washed downstream and attracted cells in that area. After many years of work the attractant was found to be cyclic AMP (Bonner *et al.*, 1969).

A variety of aggregateless strains have been isolated in several laboratories. Bonner *et al.* (1969) have shown that these can be subdivided into those that fail to excrete cAMP and those which fail to respond to cAMP. Other strains appear to synthesize cAMP and respond to it normally but are nonetheless aggregateless. This latter class probably includes those strains carrying lesions in aggregation factors not yet defined.

The related species, *Polysphondylium pallidum*, also secretes cAMP but does not respond chemotactically to this agent (Konijn *et al.*, 1969a). This may account for the observation that *D. discoideum* and *P. pallidum* will aggregate independently from populations made up of the two species.

Shortly after the environment is depleted of nutrients, the cells excrete cAMP into the medium (Fig. 5.2) (Bonner *et al.*, 1969; Malkinson and Ashworth, 1972). This could result from an increase in the activity of the synthetic enzyme, adenyl cyclase, or an inhibition of the degradative enzyme, phosphodiesterase, or from a modification of the flow pattern such that excretion is stimulated. Adenyl cyclase has been found to remain at about the same specific activity throughout development and so cannot directly account for the dramatic increase in cAMP during aggregation (Rossomando and Sussman, 1972, 1973).

Changes in cAMP phosphodiesterase could account for the kinetics of secretion of cAMP if the enzyme were active before the aggregation

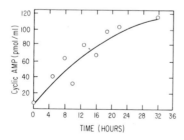

Fig. 5.2. Changes in extracellular cAMP concentration during development of *D. discoideum* Ax2 in the absence of added salts. cAMP was assayed by isotope dilution of ³H-cAMP binding to protein kinase (Malkinson and Ashworth, 1973).

stage and subsequently became inhibited or inactivated. In fact, growing cells secrete a phosphodiesterase with a K_m of 15 μM for cAMP into the medium during the last few hours of growth on bacteria before the complete depletion of nutrients from the medium (Gerisch *et al.*, 1972; Reidel and Gerisch, 1971; Reidel *et al.*, 1972; Pannbacker and Bravard, 1972). This enzyme effectively removes cAMP from the environment and keeps cells from prematurely aggregating. When the nutrients are completely exhausted, the amoebae secrete a protein which interacts with the phosphodiesterase and inhibits its activity (Gerisch *et al.*, 1972). The inhibitor is heat stable and can be purified from aggregating cells. By interaction of the inhibitor with the enzyme, the total activity of phosphodiesterase in the environment drops drastically and the concentration of cAMP can build up (Fig. 5.3). Mutant strains which fail to secrete the inhibitor are unable to aggregate (Reidel *et al.*, 1973). Other aggregateless strains overproduce phosphodiesterase. It is possible that the excess phosphodiesterase in these strains destroys the gradient in cAMP and directly blocks aggregation, but we cannot rule out the possibility that the mutation results in pleiotropic effects which independently block both aggregation and secretion of the inhibitor. Other lines of evidence directly implicate the phosphodiesterase–inhibitor system in chemotaxis. When the extracellular phosphodiesterase is partially inhibited by reducing agents at a stage before the secretion of the protein inhibitor, chemotaxis to cAMP is potentiated (Pannbacker and Bravard, 1972). Likewise, mutant strains which secrete reduced amounts of phosphodiesterase aggregate over greatly extended areas (Riedel *et al.*, 1973). However, if amoebae are treated with antibodies prepared against the phosphodiesterase, aggregation is completely blocked (Goidl *et al.*, 1972). This appears to be a contradictory result since inhibition by

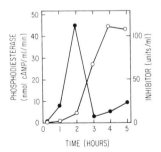

Fig. 5.3. Phosphodiesterase (PD) and PD-inhibitor concentration in *D. discoideum* Ax2. Growing cells were washed free from medium and suspended at 5×10^7 cells/ml in 0.017 M phosphate buffer pH 6.0; ●——● phosphodiesterase activity; ○——○ inhibitor activity. One unit of inhibitor inactivates 1 unit of phosphodiesterase (Reidel *et al.*, 1973).

the natural inhibitor does not impair aggregation. A possible explanation is that phosphodiesterase may play 2 separate roles in the chemotactic aggregation of *D. discoideum*. An enzyme with properties similar to those of the excreted phosphodiesterase accumulates in the cell membranes of aggregating amoebae (Chassy, 1972; Malchow *et al.*, 1972; Pannbacker and Bravard, 1972). It has been suggested that in this state the enzyme may bind cAMP and direct cell movement. Coating the enzyme with antibody would therefore inactivate the sensing mechanism. Mutant strains that fail to acquire the membrane-bound phosphodiesterase are unable to aggregate (Malchow *et al.*, 1972). This cell-bound enzyme is insensitive to the secreted inhibitor protein and thus may function to insure that the cells will be in a gradient of cAMP produced by neighboring cells.

It is clear that cells of *D. discoideum* respond to cAMP and move toward regions of higher concentration (Konijn *et al.*, 1968). Some recent evidence has implicated a role for Ca^{2+} ions in this response (Mason *et al.*, 1971; Chi and Francis, 1971). It was found that the addition of exogenous cAMP to *D. discoideum* leads to the rapid release of Ca^{2+} to the environment. There appears to be a requirement for Ca^{2+} ions at a concentration above 10^{-5} M for normal aggregation. At concentrations above 1 mM, Ca^{2+} inhibits cAMP secretion. Exogenous cAMP at 10^{-4} M also inhibits secretion of cAMP (Mason *et al.*, 1971). It has been pointed out that calcium ions have been shown to stimulate actin-myosin contraction and thus could affect amoeboid movement by local interaction with similar contractile proteins.

Periodic Movement

A curious phenomenon observed in time-lapse microcinematography of dense aggregating populations of *D. discoideum* is that chemotaxis to the center occurs in outward propagating concentric waves which have a period of about 5 minutes (Gerisch, 1968; Robertson *et al.*, 1971). These waves are thought to result from production of cAMP by cells about 15 seconds after exposure to an increase in cAMP concentration. A portion of the adenyl cyclase of *D. discoideum* is localized in the plasma membrane of the cells and thus is in a position to rapidly respond to changes in extracellular cAMP concentrations (Rossomando, 1974). For about 5 minutes after responding to cAMP the cells are refractory to further stimulation by the nucleotide. Unfortunately, direct determination of pulsation in cAMP production has been impossible to perform. Nevertheless, the pulsatile aggregation has been able to account for the dramatic streaming of *D. discoideum* (Schaffer, 1957a; Robertson, 1972).

It has been argued that were cAMP secreted continuously, amoebae would migrate directly to the point of highest concentration and would not form streams. Pulsatile secretion of cAMP and a refractory period may insure that amoebae will respond chemotactically to cells in the immediate neighborhood. Since the signal is relayed outward from cell to cell, the aggregation territory can enclose a large area. This argument suggests that parameters which affect the size of the aggregation territory, both genetic and environmental (M. Sussman, 1956b; Konijn and Raper, 1961), may modify the frequency of pulsation or chemotactic sensitivity. In support of this idea, both Gerisch (1971) and Durston (1974) have isolated mutants of *D. discoideum* which secrete cAMP continuously without periodicity, and have shown that these strains fail to aggregate in streams. Moreover, the aggregation territory of these strains is greatly reduced.

Robertson *et al.* (1972) have been able to entrain amoebae of *D. discoideum* to pulses of cAMP released periodically at intervals of about 5 minutes from a micropipette. During the first 4 hours after the initiation of development, the cells failed to respond, but thereafter the amoebae could be seen to move toward the cAMP source, usually within 30 seconds after the release of the pulse of cAMP. Cells up to 100 μm from the micropipette were seen to respond. Cells that had developed for 4 hours responded chemotactically but did not propagate the signal. About 2 hours later, however, signal propagation could be observed. The rate of propagation was 42 μm/second, close to that observed in natural aggregation (Gerisch, 1968). Finally, about 8 hours after the initiation of development, spontaneous signaling occurred in various parts of the population. These observations indicated that there is an interphase period following the removal of exogenous nutrients during which the cAMP detection and movement-response mechanisms are developed. A few hours later the signal relay and spontaneous signaling mechanisms become fully functional. After about 8–9 hours of development, well-defined streams arise in the population indicating that intercellular contacts have increased and the cells have increased their mutual cohesion.

For many years it was thought that spontaneous signaling required the presence of a specialized differentiated cell (I-cell) in the population (Sussman, 1956b). Time-lapse microcinematography has shown that such a cell is not normally required for aggregation of *D. discoideum* (Gerisch, 1968). Under most conditions, cells aggregate to one or more cells which raise the local concentration of cAMP over the background. In some instances an exceptionally large cell becomes the first to attract its neighbors. It has been suggested that these large cells may be diploids or aneuploids found at a low frequency in such populations (Ashworth and Sackin, 1969).

Cohesion of Developing Cells

Shortly after the initiation of development, cells agglutinate and form clumps if swirled in salt solutions (Gerisch, 1961b; Born and Garrod, 1968). Under these conditions 1 mM EDTA prevents clumping and blocks aggregation (Gerisch, 1961b; Gingell and Garrod, 1969). Later in aggregation, about 8–12 hours after the cells are washed free of exogenous nutrients, cells swirled in phosphate buffer will agglutinate even in the presence of 1 mM EDTA. However, clumping of these cells can be prevented by concentrations of EDTA from 1 to 2×10^{-2} M (Gerisch, 1961b).

Cohesion is stimulated by calcium ions which may act as intercellular bridges (Born and Garrod, 1968). Thus, EDTA may chelate surface bound Ca^{2+} and in this way prevent cohesion. Alternatively, cell bound Ca^{2+} might be removed by the EDTA, leaving the cells with a net negative surface change which would tend to result in repulsion of the cells. However, it has been shown that the net surface electrostatic potential as well as the zeta potential of aggregating cells are unchanged by EDTA (Gingell *et al.*, 1969; Gingell and Garrod, 1969; Lee, 1972a,b). It is presently unclear how EDTA interferes with cohesion.

Both the zeta potential and the surface charge density decrease spontaneously during the first few hours following the initiation of development. Garrod and Gingell (1970) have suggested that the electrostatic repulsive forces between cells may be a limiting factor in determining adhesive stability. A decrease in the surface charge would then be expected to promote cohesion. However, these authors also considered the possibility that the change in surface charge might simply reflect the appearance of new surface components as the cells become cohesive. Previous work has shown that new antigenic determinants accumulate on the cell surface of aggregating cells and that these components do not arise in aggregateless mutant strains (Gregg, 1956; Gerisch, 1968; Gregg and Trygstad, 1958). It now seems likely that the increase in cohesion is mediated by specific surface proteins which may inadvertantly alter the surface charge.

Beug *et al.* (1973a,b) have been able to prepare antisera specific to cell surface components of aggregated cells of *D. discoideum*. The sera contain antibodies to many cell surface determinants, but those directed at the components found before the initiation of development

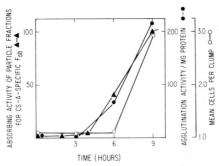

Fig. 5.4. Developmental kinetics of agglutination factor and specific surface antigen. Amoebae of strain NC-4 were either allowed to develop on filters (for determination of absorbing activity) or suspended in buffer (for determination of absorbing activity). Cohesion was assayed by measuring the mean size of clumps formed in 10^{-2} M EDTA pH 6.5 in 30 minutes in roller tubes using a Coulter Counter. Agglutination factor was assayed using fixed sheep red blood cells. Absorbing activity was assayed by measuring the ability of cell extracts to inhibit the activity of Fab fragments of antibody to specific cell surface determinants (Rosen *et al.*, 1973 and Beug *et al.*, 1973a,b).

can be removed by adsorption to vegetative cells. Antibodies in the absorbed sera have been purified and shown to agglutinate developing but not vegetative cells (Beug *et al.*, 1973a). The specific surface components recognized by the antibodies accumulate about 9 hours after the initiation of development, at about the same time that the cells can be observed to become cohesive in the presence of EDTA (Fig. 5.4). Recently, accurate methods to quantitate cohesion among cells of *D. discoideum* have been developed that involve measurement of the light scattering properties of a population of cells slowly rotated while suspended in phosphate buffer (Born and Garrod, 1968; Beug and Gerisch, 1972). This method has shown that there is a dramatic increase in cohesion among the cells between 7 and 10 hours of development (Fig. 5.4). Mutant strains that fail to aggregate do not increase in cohesiveness by this test. The test gives a meaningful estimation of the changes in cohesion associated with development. The coincidence of the time of appearance of the surface antigen and increased cohesion indicate a close relationship.

Univalent Fab fragments have been prepared by treatment of the antibodies specific to aggregation with proteolytic enzymes (Beug *et al.*, 1973a,b). These Fab fragments do not agglutinate the amoebae, but block rotation-mediated cohesion of developed cells. It is thought that these Fab fragments bind to cell surface sites which are essential for intercellular contacts. Fab fragments of antibodies prepared against undeveloped cells will bind to growing cells and block the weaker, EDTA-sensitive cell cohesion, but do not affect the strong mutual cohesion that is seen in cells allowed to develop for 9 hours. Thus, attachment of nonspecific Fab fragments does not block the developmentally controlled cohesion while specific Fab fragments are effective in inhibiting this process. Further understanding of the mechanisms of aggregation and differential cohesion will depend on elucidating the structural and biochemical nature of the antigenic sites associated with cohesion.

An independent approach to elucidating the mechanism of cohesion is to chemically dissect the system in hopes of finding a specific cohesion factor and associated binding sites. This approach is based on the assumption that cell contacts are held together by multivalent proteins with affinity to cell surface sites. Cohesive molecules of this type have been demonstrated in the species-specific cohesion of sponge cells but have been difficult to show in other systems. Recently, Rosen *et al.* (1973) have isolated a protein of 100,000 daltons from aggregating cells of *D. discoideum* which will agglutinate fixed red blood cells. The molecule is composed of 4 identical subunits of 25,000 daltons and accumulates at the same time the cohesion of the cells is seen to increase (Fig. 5.4). The factor is present on the surface of the cells as shown by the fact that rosettes of bound red blood cells will form around aggregatting cells. Rosette formation, as well as agglutination, can be inhibited by *N*-acetylgalactosamine, but not by *N*-acetylglucosamine, suggesting that the binding sites may contain *N*-acetylgalactosamine.

So far it has been difficult to study agglutination of *D. discoideum* cells themselves by the factor since conditions which block spontaneous agglutination also appear to block the action of the factor. It has not yet been conclusively proven that this factor actually accounts for cohesion in this system.

The accumulation of the factor was determined in a series of morphological mutants which were blocked in the initial steps of aggregation and never acquired cohesiveness. Two of the 4 mutant strains tested did not accumulate the factor while 2 others contained the factor (Rosen and Loomis, unpublished). It is not clear whether the primary block in those strains that failed to accumulate the factor was the lack of the factor itself or resulted from an earlier event necessary for factor accumulation. Only the isolation of temperature-sensitive aggregation mutants in which the factor itself is thermolabile can directly indicate its essential function.

Time-lapse microcinematography has shown that amoebae adhere to each other chiefly by their ends when they enter an aggregation stream (Schaffer, 1964a) (Fig. 5.1). It is not clear whether this observation is due to the plastic deformation of the cells as they attach to the moving stream of cells, or due to an initial polarity of the cells and localization of cohesive sites. Based on a variety of observations, Beug *et al.* (1973a) have suggested that the cohesion sites which are inserted into the cell surface during development are localized at the ends of the elongated cells. Using fluorescein-labeled Fab fragments of the antibody directed against the aggregation-specific determinants, Beug *et al.* (1973b) found that the sites were evenly distributed over the cell surface and were not localized to the ends. They could also calculate from the binding of ^{3}H-labeled Fab fragments that there are about 3×10^5 antigenic sites on the surface of aggregating cells and that they make up at least 2% of the total surface area. During the first few hours after the initiation of development a considerable portion of the cell surface is modified.

Several other changes have been observed in *D. discoideum* between 6 and 12 hours of development. Aldrich and Gregg (1973) have found that the ultrastructure of the membrane changes during this period. Characteristic projections seen by freeze-etch electron microscopy in the split lipid bilayers of the plasma membrane appear to coalesce to form more prominent projections about 8 hours after the initiation of development. Aldrich and Gregg speculated that this change may account for the increase in cohesion seen at about this time in development. The modification of the membrane occurs precociously in cells incubated in the presence of cAMP (Gregg and Nesom, 1973). Cyclic AMP may not only elicit chemotactic responses but also induce cellular changes leading to increased cohesion of the cells.

Although many of the steps in aggregation are only partially characterized we can already see a sequence of events which appears to account for this dramatic change in cell behavior. As the food source becomes limiting the cells secrete a phosphodiesterase which keeps the external concentration of cAMP low. A few hours later a specific inhibitor of phosphodiesterase is secreted and a gradient in cAMP can be established. A few cells spontaneously emit cAMP and surrounding cells respond chemotactically by moving toward higher concentrations of the attractant. The responding cells are also stimulated to emit cAMP. After emitting a burst of cAMP the cells require a period of 5 minutes before releasing further cAMP. The refractory period, signal relay, and subsequent entrainment result in a definite periodicity and polarity to aggregation, and the formation of well-defined streams. As cells accumulate at the centers, they synthesize a specific protein which is inserted into the plasma membrane and results in increased cellular cohesion. The cells then form tight aggregates which soon transform into pseudoplasmodia.

Chapter 6

Pseudoplasmodial Stage

When Brefeld (1869) first identified and described a member of the Acrasiales over a hundred years ago, he considered it to be closely related to the Myxomycetes which aggregate and fuse to form multinucleate plasmodia. This error has generated considerable confusion up to the present since the 2 groups of organisms still share the common name "slime molds." Van Tieghem (1880) was the first to show that members of the Acrasiales remain as individual uninucleated cells during aggregation and throughout the life cycle. To emphasize the cellular nature of the Acrasiales aggregates he referred to them as "pseudoplasmodia." The term is a negative one, merely correcting a false observation and thus has been replaced in some of the recent literature from English laboratories with the term "grex" (Shaffer, 1962). I have chosen to continue usage of the original term.

The pseudoplasmoidal stage is an obligatory step in the morphogenesis of *D. discoideum.* It is characterized by the integration of aggregating cells in a fingerlike structure surrounded by a cellulose sheath (Fig. 6.1). Pseudoplasmodia can be formed from as few as 100 or as many

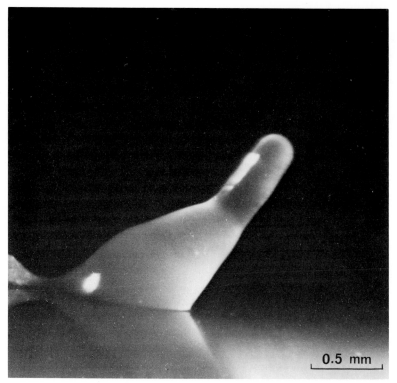

Fig. 6.1. An early pseudoplasmodium. Aggregated cells become surrounded by a surface sheath and rise off the substratum.

as 100,000 cells (Raper, 1940b). Although pseudoplasmodia often migrate over considerable distances, migration is not essential for subsequent morphogenesis.

Integration

The number of cells entering a pseudoplasmodium is determined by the number of cells in the aggregation territory (M. Sussman, 1956b; Bonner and Dodd, 1962a). Sparse cultures will form small pseudoplasmodia containing only a few hundred cells while dense populations will give rise to pseudoplasmodia containing up to 100,000 cells which can be up to several millimeters in length. The entry of cells into a pseudoplasmodium continues until the surface sheath can contact the support and effectively seal off the mass of cells from any late comers. The upper limit on size is probably determined by cohesion of the

cells, shear forces generated by cell movement, and properties of the
sheath. When very large aggregates are formed they often subdivide
and construct several separate pseudoplasmodia which then proceed
independently through morphogenesis. Other times, large pseudoplasmodia split into smaller ones during migration.

The sheath components are secreted by the cells shortly after the
initiation of development and throughout the pseudoplasmodial stage
(Freeze and Loomis, in press). Once incorporated into the sheath they
are highly insoluble even in 8 *M* urea and 2% sodium dodecyl sulfate
and can be separated from other cellular components on this basis. About
70% of the sheath is composed of polysaccharides containing glucose,
mannose, *N*-acetylglucosamine, and galactose and a few other sugars
in lower proportions. The attached proteins are fairly small and contain
an exceptionally high content of asparagine. About 85% of the polysaccharide component appears to be cellulose by a variety of techniques.
Moreover, Hohl and Jehli (1973) have reported observing cellulose fibers
in the sheath left behind during migration. The presence of cellulose
fibers may account for the surprising strength of the sheath.

After the initial formation of fingerlike structures, pseudoplasmodia
bend over and become horizontal (Fig. 6.2). In this mode they can
migrate for a period of days, covering tens of centimeters. The conditions
which favor migration are those of high humidity and low ionic concentration (Raper, 1940b; Newell *et al.*, 1969). The presence of any of
a variety of monovalent or divalent ions inhibits migration and leads
to culmination at the place of aggregation. The mechanism for this be-

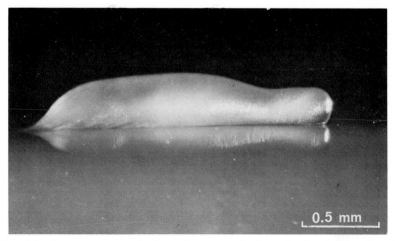

0.5 mm

Fig. 6.2. A migrating pseudoplasmodium. Pseudoplasmodia bend over onto the
substratum and migrate horizontally.

havioral control is unknown, but one can consider that these environ-mental conditions may affect amoeboid movement or the physical proper-ties of the sheath. When development occurs in the absence of migration, the cells nevertheless move relative to the sheath which then accumulates as a skirt around the base (Shaffer, 1965c). This "treading in place" is thought to be important for the regulation of tissue proportions as it ensures that the apical cells are surrounded by recently deposited sheath material as is discussed in Chapter 13.

Polarity of Migration

During pseudoplasmodial migration, new sheath material is produced which is extended only at the tip as the cells move forward while the previously formed sheath is left behind as a trail (Francis, 1962). Except at the anterior tip, the sheath itself is immobile while the cells move in relation to it. The sheath extends over the tip and thus must be newly formed as the pseudoplasmodium advances (Fig. 6.3). Pseudoplasmodia migrate with a strict polarity and are never seen to move sideways or backward but always in the direction of the anterior tip. This is probably a consequence of being enclosed in the surface sheath which restricts movement to the direction of the tube (Garrod, 1969a; Loomis, 1972). Anterior movement is possible only because newly formed sheath is more fluid and can be extended before it matures (Francis, 1962). Posterior movement is blocked by the collapse of the sheath which is highly cohesive. The crucial role played by the sheath in migrational polarity was demonstrated by microsurgically removing the sheath from the posterior end and observing a reversal of migrational polarity in a significant number of cases (Loomis, 1972). Pseudoplasmodial cells are able to move in any direction, but only if they are presented with a surface free of preformed sheath.

The rate of migration is dependent on the size of the pseudoplasmo-dium, being greater in large pseudoplasmodia than it is in small ones (Bonner et al., 1949). It has been suggested that the motive force is determined by the total number of cells and that the frictional drag is dependent on the surface area (Francis, 1962). The net rate of migra-tion is then determined by the ratio of total number of cells to total area. Since the number of cells determines the volume, the rate will be proportional to the radius.

It is not clear what limits the rate of migration in small pseudoplas-modia but we can consider the possibility that it is sheath formation. If the rate of synthesis of the precursors of the sheath is dependent

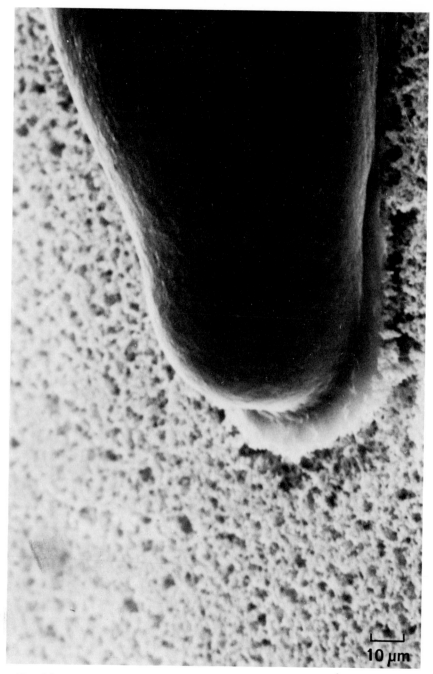

Fig. 6.3. Anterior of a migrating pseudoplasmodium. The surface sheath appears to be a smooth uniform covering which extends over the tip and flows somewhat onto the substratum, in this case a Millipore filter (SEM).

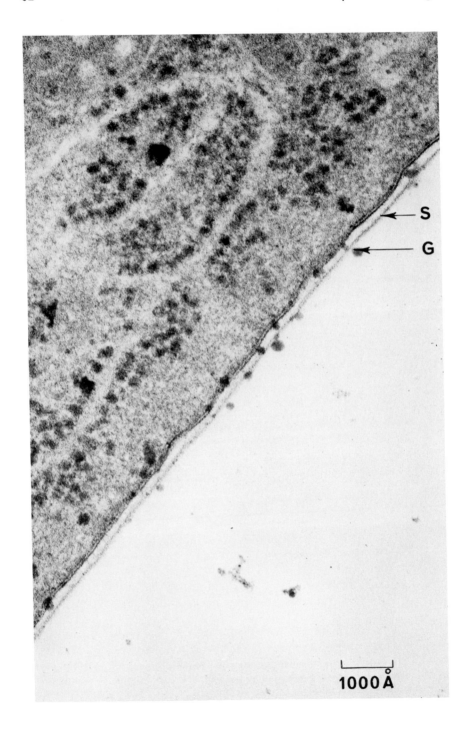

on the total number of cells, the production of sheath will be a function of the surface-to-volume ratio. Since the axial ratios are similar over a considerable range in pseudoplasmodial size, large pseudoplasmodia will have a lower surface-to-volume ratio and so should be able to deposit sheath at a greater rate. If the increase in sheath production subsequently allows an increased rate of migration, we would expect the sheath thickness to be size invariant. This was directly tested in a series of migrating pseudoplasmodia which varied by a factor of more than 200 in number of cells (Farnsworth and Loomis, 1975). The sheath surrounding the pseudoplasmodia can be clearly visualized by transmission electron microscopy (Fig. 6.4). It completely surrounds the outside of pseudoplasmodia and appears to polymerize only on the surface (Fig. 6.5). The thickness increases along the length as was predicted from the observation that subunit incorporation occurs throughout the length of pseudoplasmodia (Loomis, 1972). Moreover, it can be seen that the thickness varies little among pseudoplasmodia of quite different sizes (Fig. 6.6). Large pseudoplasmodia which migrate more rapidly seem to produce more sheath.

Phototaxis

Pseudoplasmodia of *Dictyostelium* respond to light by migrating toward the source (Raper, 1940b; Bonner *et al.*, 1950). This behavior increases the chances that pseudoplasmodia which form within the natural habitat of the forest detritus will migrate to the surface where spore dispersal can be more effective.

Various aspects of the mechanism of phototaxis have been elucidated recently. As in any photoresponsive system, light must be able to excite a molecule which can trigger the biochemical processes which lead to the response. The response in this case consists of changing the direction of migration. As a first step in defining the system, the action spectrum for the turning reaction was measured (Poff *et al.*, 1973). By placing a uniform population of pseudoplasmodia between a pair of lights of different wavelengths held at right angles to each other, the preferred wavelength could be determined by measuring the angle of migration relative to the light sources. With lights of identical wavelength pseudo-

---

Fig. 6.4. Surface sheath. The thin extracellular layer 150 Å thick in this section is seen to be a continuous sheath (S). In some sections the sheath appears trilaminar. Small granules (G) 150 Å in diameter are seen attached to the sheath in all specimens. The sample was fixed with glutaraldehyde and osmium tetroxide in the presence of ruthenium red (TEM).

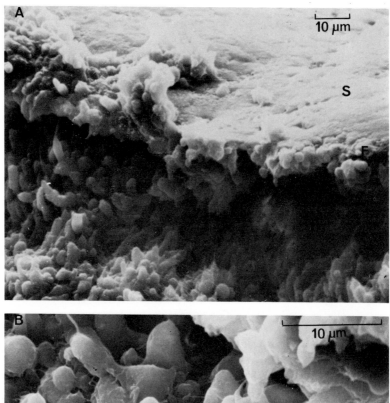

Fig. 6.5. Surface of a fractured pseudoplasmodium. (A) The sheath (S) appears as a smooth covering up to the edge of the fracture (F). (B) Higher magnification of the cells exposed by the fracture reveals the loose packing of the cells and the lack of interstitial material similar to the surface sheath. The fibers connecting the cells may be poorly preserved cellular projections or sheath material which has not fully polymerized within the pseudoplasmodium (SEM).

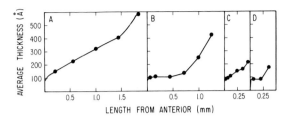

Fig. 6.6. Thickness of the surface sheath around pseudoplasmodia of different sizes. Pseudoplasmodia migrating phototactically were fixed in glutaraldehyde and ruthenium red and prepared for electron microscopy. Transverse sections were taken along the length and examined at 60,000× magnification. The sheath was traced from negatives and its average width was determined. The thickness is plotted at the position of the section.

plasmodia migrated on a line bisecting the angle between the lights, suggesting that migration in this case resulted from a balance of response to each source. The pseudoplasmodia were found to be most responsive to blue light at 430 nm and to green light in the range of 550–590 nm (Fig. 6.7). Using a different technique with lower resolution, Francis (1964) also found the pseudoplasmodia to have maximal phototactic responses to wavelengths in these two ranges. Irradiation at 560 nm results in a rapid increase in absorption of a pigment measured at 411 nm (Fig. 6.8) (Poff *et al.*, 1973). The light-induced absorption change decays with a half-life of about 7 seconds when the irradiation is stopped (Fig. 6.8). The response could be observed in cell-free preparations and was localized in the mitochondrial fraction. The photopigment,

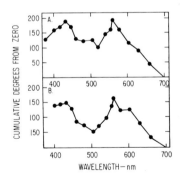

Fig. 6.7. Action spectrum for phototactic migration. Relative response of phototactic migration of pseudoplasmodia to constant energies of monochromatic light: (A) 1 μW/cm²; (B) 10 μW/cm². The relative response is the angle of migration away from the bisect of two orthogonal light sources with adjacent wavelengths. The maximal response is 45°. Cumulative values starting at 700 nm are plotted against wavelength (Poff *et al.*, 1973).

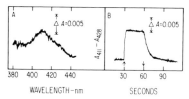

Fig. 6.8. Light-induced absorbance change. Axenically grown cells were broken in a Dounce homogenizer and fragments removed by centrifugation at 1000 g for 5 minutes. The supernatant was then centrifuged at 12,000 g for 30 minutes. All activity was found in the pellet. (A) Light-minus-dark difference spectrum due to irradiation with 560 nm light on; (B) kinetics of the light-on and light-off response measured as the absorbance difference between 411 nm and 428 nm (Poff *et al.*, 1973).

phototaxin, has been solubilized and purified to greater than 90% homogeneity (Poff *et al.*, 1974). It is a heme protein of 240,000 molecular weight which undergoes a reversible photooxidation when irradiated with light of the appropriate wavelength. Both the action spectrum of photooxidation and the absorption spectrum of the molecule (Fig. 6.9) are similar to the action spectrum of migration (Fig. 6.7). Phototaxin, thus appears to be the primary photopigment of the light response.

Once the light is absorbed, the oxidized phototaxin must direct reactions which result in the turning of pseudoplasmodium. Since the molecule appears to be localized in mitochondria, the photoactivation most likely modifies reactions localized in these organelles. Exactly how the modifications affect migrational direction is still an open question.

Phototaxis is known to be mediated by the "lens effect" first described in Phycomycetes by Blaauw (1909). Incident light is focused by the convex surface of the pseudoplasmodium onto the distal side (Francis, 1964; Poff and Loomis, 1973). As shown by the use of vertical microdot

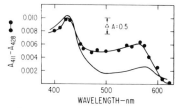

Fig. 6.9. Absorption and action spectra of phototaxin. Purified phototaxin (0.5 ml) was reduced with dithionite and measured in the presence of 0.5 gm $CaCO_3$. Sample thickness was 0.3 cm (solid line). The action spectrum of similar material was determined by measuring the change in absorption at 411 nm relative to 428 after irradiation with light of constant energy at various wavelengths (●———●) (Poff *et al.*, 1974; Poff and Bulter, 1974).

illumination along the pseudoplasmodial axis, the organism is only responsive at the anterior tip (Fig. 6.10) (Poff and Loomis, 1973). When the vertical microdot is placed so as to illuminate cells on one side of the tip, these cells can be seen to be stimulated within 15 seconds. They surge ahead and then turn toward the dark side as the cohesive forces in the pseudoplasmodium pull them around. Even after the light is shut off the cells keep turning for several minutes. The net result is that the pseudoplasmodium turns away from the illuminated side and proceeds on an altered course. These observations suggested that irradiated pseudoplasmodia should migrate more rapidly than unirradiated pseudoplasmodia. Raper (1940b) had been unable to observe any effect of light on migration rate, but he used a population of pseudoplasmodia with widely varying migration rates. When the problem was reinvestigated using a population of pseudoplasmodia of uniform size and migration rate, it could be seen that light stimulated the rate of migration up to 75% (Poff and Loomis, 1973). We can now propose a model for control of phototaxis in *D. discoideum:* light is focused onto distal cells

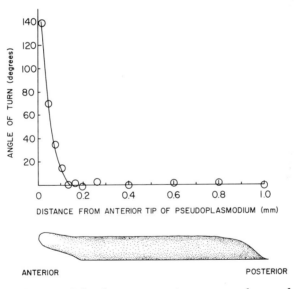

Fig. 6.10. Localization of the phototactic turning response of a pseudoplasmodium. The pseudoplasmodium was irradiated on one side for 5 minutes and permitted to complete its turn in darkness for 15 minutes following the irradiation period. The angle of turn induced by light was plotted vs. the portion of the pseudoplasmodium irradiated. The diameter of the irradiating microbeam is represented by the diameter of the data points. Each data point is the average of two separate determinations (Poff and Loomis, 1973).

by the lens effect of the pseudoplasmodia and is preferentially absorbed by phototaxin in those cells; photooxidation of the phototaxin results in an increase in migration rate in that region and subsequent turning of the pseudoplasmodium toward the light as it pivots on the relatively slowly advancing proximal cells. This mechanism is similar to turning a tracked vehicle by increasing the speed of the outside track. Once the pseudoplasmodium is migrating directly at the light, both sides become equally stimulated and further turning ceases. Thus, phototaxis results from the effect of light on a rate-limiting step in migration. This step could be in the rate of amoeboid motion of the individual cells or on some process which integrates the pseudoplasmodium.

Several investigators have tried unsuccessfully to stimulate the rate of amoeboid movement of cells dissociated from pseudoplasmodia (Francis, 1962; Samuel, 1961). The movement of single cells or small clumps of cells seems to be completely unresponsive to a variety of illumination techniques. Therefore, it is likely that phototaxis is mediated by a structure found in pseudoplasmodia but not in the cells. The surface sheath is just such a structure and may play a role in directing migration as well as controlling the rate of migration. If light stimulates sheath production over the distal side, the whole pseudoplasmodium would turn to the light.

Mutant strains have been isolated in which migration is apparently normal except that the pseudoplasmodia fail to orient to light (Loomis, 1970b). If cells of these "blind" mutants are mixed in equal number with wild-type cells, the resultant pseudoplasmodia orient to light but do so poorly, suggesting that each cell responds separately to light and cannot transmit the signal to adjacent cells. These strains, of which there are now five of the totally nonphototactic class, all contain the photopigment, phototaxin, and therefore must carry lesions in the pathway leading from the activated phototaxin to the turning response (Poff and Loomis, unpublished).

Thermotaxis

Pseudoplasmodia will orient in an environment in which there is a temperature gradient and will migrate into warmer areas (Raper, 1940b). Bonner *et al.* (1950) studied thermotaxis of *D. discoideum* in detail and found that pseudoplasmodia would orient in a field in which the temperature differential was only $0.05°C/cm$. Since small pseudoplasmodia are only about 0.1 mm wide, this means that the temperature difference between the two sides was less than $0.0005°C$. This extreme sensitivity to heat led Bonner *et al.* (1950) to suggest that phototaxis might result from differential heating of the sides of a pseudoplasmodium by the absorbed light. However, it can be calculated that the heating of the cells by blue light is insignificant at intensities to which the pseudoplasmodia respond phototactically. Moreover, the mechanisms of phototaxis and thermotaxis must differ significantly since pseudoplasmodia respond to both light and heat by turning toward the source, yet phototaxis is known to depend on the lens effect of the convex pseudoplasmodial surface. Thermotaxis cannot involve a lens effect since heat would not be focused by the cells. Thus, phototaxis results from stimulation of the distal side while thermotaxis must result from an effect of heat on the proximal side. It is possible that a temperature gradient retards movement of the proximal side by affecting the local relative humidity. A decrease in humidity might reduce the rate of amoeboid movement or might increase the restrictive properties of the surface sheath on the proximal side. Over a period of hours even a small effect on these properties could turn the direction of migration toward the source. This mechanism would be similar to turning a tracked vehicle by braking the inside track. Since individual amoebae do not orient in a temperature gradient (Bonner *et al.*, 1950), the response is probably mediated through a pseudoplasmodial function such as the integrative and directive properties of the surface sheath.

Cell Sorting in Pseudoplasmodia

The first cells to aggregate usually take up an anterior position in pseudoplasmodia and ultimately give rise to stalk cells. However, it has been suggested that under certain circumstances considerable cell sorting occurs during aggregation. If the center of an aggregate is removed and replaced by cells marked with a vital dye, the resulting pseudoplasmodia are found to be uniformly colored (Bonner and Adams, 1958). If the dye had remained associated with the central cells and these had retained their position, the resulting pseudoplasmodia would have been stained only in the anterior portions. Either considerable reshuffling of the cells occurs during aggregation or the method of staining results in diffusion of the dye throughout the cell mass.

Takeuchi (1969) reported that a spread in cell density occurs in a population of *D. discoideum* and that the densest cells have a preference for the anterior of pseudoplasmodia. These dense cells may be metabolically more active than the majority of the cells and may initiate chemotaxis before other cells, thus becoming the center of the aggregates.

It has also been found that prior growth conditions can dramatically affect the predisposition of cells to take up anterior positions in pseudoplasmodia (Leach et al., 1973). Cells taken in the stationary phase of growth take up positions anterior to those of cells in the exponential phase. It is likely that in the stationary stage the cells undergo many of the differentiations which prepare them for aggregation and that when mixed with cells from the exponential phase, the stationary stage cells initiate most of the aggregates. Likewise cells growing exponentially in a medium lacking glucose will tend to take up a position anterior to cells growing exponentially in complete medium. In this case, growth in the glucose-free medium, which is far slower than in complete medium, may result in precocious preparation for aggregation during the growth stage. It is clear that chemotaxis is not the only determining process in this bias since cells agglutinated by rolling will form pseudoplasmodia with the cells sorted out on the basis of their previous growth condition.

Leach et al. (1973) investigated whether the proclivity of the cells to a peculiar position in a pseudoplasmodia resulted from density differences caused by the different growth conditions. It was found that growth in the presence of glucose resulted in denser cells which, from Takeuchi's work, would be expected to take up anterior positions; however, they preferentially took up posterior positions. These authors suggested that growth conditions might affect the cohesive properties of cells which could result in sorting out during aggregation.

While there is little question that cell mixing can occur during aggregation, there has been conflicting data as to whether cells move relative to each other during pseudoplasmodial migration. Raper (1940b) labeled cells by growing them on the red bacteria Serratia marcescens. These cells were allowed to develop until pseudoplasmodia were formed, at which point the tips were removed and replaced by colorless cells from the tip of an unmarked pseudoplasmodium. The line of demarcation between the marked and unmarked regions was found to retain its shape and did not move during subsequent migration. Raper (1940b) described the junction between the red and colorless cells as resembling the joining of a red and a white brick wall with the cells closely opposed and only slightly interdigitated. He concluded that there was little cell mixing or sorting during pseudoplasmodial migration.

Bonner (1952) extended this study by grafting posterior cells from a vitally stained pseudoplasmodium onto the anterior of an unstained posterior portion. As migration proceeded, the dye was seen to travel toward the posterior end as a fairly well-defined band. Bonner concluded that more rapidly moving cells were moving past the vitally stained cells and taking up an anterior position. Cell mixing and sorting were assumed to be occurring. Cell sorting was considered to result from differential migration rates of the individual amoebae and forms the basis of Bonner's analysis of tissue proportioning in *D. discoideum* (Bonner, 1957). However, on the basis of apparent intercellular adhesion in pseudoplasmodia, Francis (1962) showed that the forces of cohesion exceeded those of amoeboid movement by at least an order of magnitude in *D. discoideum*. He concluded little or no cell mixing in possible in this system.

The problem was reinvestigated by Farnsworth and Wolpert (1971), who labeled cells internally with particle-bound acridine orange. The technique is similar to that of Raper but of greater resolution. In agreement with Raper's observations they found that in no case did the line of demarcation fade between any grafted piece and the host. The results indicated that less than 5% of the cells move further than 2 cells from the line of demarcation no matter what the position of origin of the grafted piece or the position into which it was grafted. They also reinvestigated Bonner's vital staining method which had given opposing results and discovered that the dye itself moved from one cell to the next. It is well known in several systems that the vital stains are often bound to extracellular components.

Thus, it seems that considerable mixing can occur during aggregation before the cohesive system is fully developed, but that after this stage the cells remain in position within pseudoplasmodia and respond to the signals for subsequent differentiation during culmination.

Chapter 7

Culmination Stage

Stalk Formation

At the end of the pseudoplasmodial stage, the tip becomes uppermost (Fig. 7.1) and the cells undergo the terminal differentiations which result in the formation of a fruiting body. During this stage considerable morphopoietic movement results in reversal of the relative positions of anterior and posterior cells. The conditions which favor culmination are poorly understood but appear to be those which tend to inhibit migration. Lower humidity, higher temperatures, and overhead light are effective in inducing culmination (Raper, 1940b; Newell et al., 1969). Overhead light elicits an attempt at vertical phototaxis which is stymied by gravity as the cell mass rises off the support and results in an end to migration. The cessation of migration may then trigger culmination.

When culmination is initiated, a cellulose-containing ring is deposited at the neck where the tip joins the main cell mass (Raper and Fennel, 1952). In both large and small pseudoplasmodia this ring is formed by the peripheral cells in this region. It has been suggested that attach-

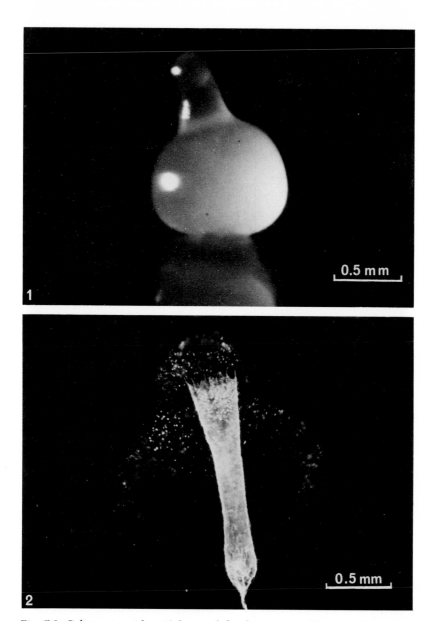

Fig. 7.1. Culmination. After 18 hours of development on 2% agar, migration stops and the anterior cells become uppermost.

Fig. 7.2. Stalk formation. The specimen seen in Fig. 7.1 was crushed and treated with 0.1% calcofluor white ST, which causes cellulose to fluoresce under the conditions used. When viewed in ultraviolet light, the stalk sheath can be seen to extend from the tip to the substratum within the culminating mass of cells. Cells near the funnel-shaped upper end fluoresce brightly and are presumably producing considerable amounts of cellulose. They may have been disturbed from their original position during specimen preparation.

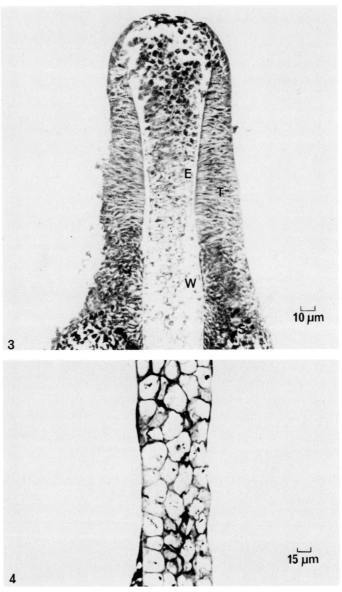

Fig. 7.3. Entry of cells into a stalk. A 1 μm section was taken through the tip of a rising stalk and stained with toluidine blue. The cells in the tip (T) are closely packed and aligned. Cells which will form spores (S) are less closely packed. When the cells enter the funnel of the stalk sheath, they expand to form larger cells (E) and then vacuolize and lay down cellulose walls (W). (See also Figs. 7.4 and 7.5.)

Fig. 7.4. Mature stalk. A 1 μm thick longitudinal section was taken through a mature stalk and stained with toluidine blue. The cells have transformed into rigid, angular bodies containing the debris of dead cells.

ment of the cells to the surface sheath is a prerequisite for the secretion of cellulose; however, no mechanism for this requirement has been presented (George, 1968). The ring is extended and forms a tubular sheath extending through the cell mass (Fig. 7.2). Cells included within this stalk sheath vacuolize and expand in volume 3- to 4-fold. Cell–cell adhesion may draw some of the cells down with the descending stalk, giving rise to a concave surface around the central stalk region and a morphological shape referred to as a "Mexican hat" (see Fig. 1.7 [18]).

When the elongating stalk contacts the solid support, further expansion is restricted to the apical axis and all but a few of the cells rise off the support on the lengthening stalk (Fig. 7.3). A few posterior cells vacuolize at the base giving rise to the basal disk characteristic of *D. discoideum*. As the stalk extends, anterior cells move into the funnel of the stalk sheath where they vacuolize and secrete cellulose to form angular walls (Fig. 7.4). The stalk tapers as it rises from the basal disk. There seems to be a correlation between the diameter of the stalk sheath and the size of the sorus at the apex (Raper and Fennel, 1952). Since the rising sorus is continuously decreasing in size as cells enter the stalk, the result is a gradual decrease in the stalk diameter. Cells which are near the base of the stalk vacuolize and undergo autodegradation while differentiating into hard walled support cells which are randomly oriented. Those cells which enter the stalk higher up vacuolize, take up a transverse orientation, and lay down rigid walls. Near the top of small stalks where the diameter is less than that of a fully vacuolized stalk cell, the cells become aligned within the sheath with their long axis parallel to that of the fruiting body (Raper and Fennel, 1952). It would appear that the cells swell when entrapped in a stalk sheath and fill the space available. When the stalk diameter does not restrict the swelling, the cells take up random positions, but when the stalk diameter limits expansion of the cells, the orientation is more regular. Once the stalk cells have vacuolized they are no longer viable (Wittingham and Raper, 1960). The stalk is constructed chiefly of cellulose and consists of the original stalk sheath and an inner matrix of angular cellulose walls formed by the included cells (Fig. 7.5).

Spore Formation

Encapsulation of spore cells begins in the periphery of the rising mass of cells (Raper and Fennel, 1952). As culmination proceeds more and more of these initially posterior cells encapsulate into heavy walled spores with the cytological changes seen in Fig. 7.6. The end of culmination is marked by the encapsulation of all cells which have not entered the stalk.

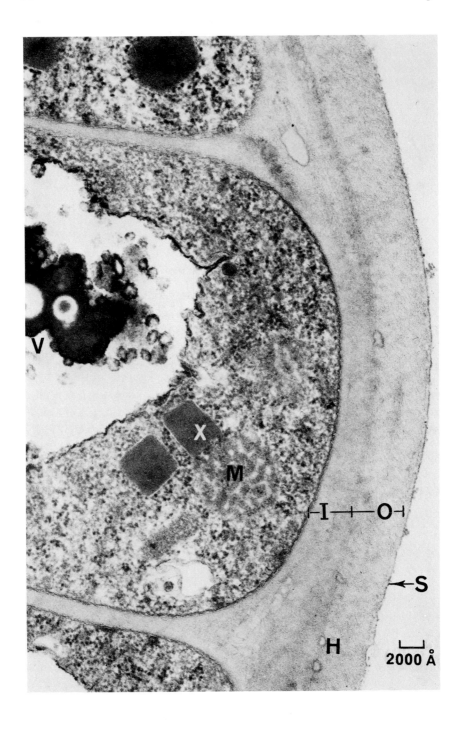

The spore wall is a multilayer structure containing both protein and cellulose (Fig. 7.7). It appears to be formed in part from organelles termed "prespore vesicles" (Maeda and Takeuchi, 1969) (Fig. 7.8). These vesicles are found in the posterior cells of pseudoplasmodia (Hohl and Hamamoto, 1969b; Gregg and Badman, 1970; Takeuchi, 1972). It has been suggested that these heavy walled vesicles fuse with the plasma membrane of spore cells and contribute their contents to the outer wall (Maeda and Takeuchi, 1969). Antiserum prepared against spores of *D. mucoroides* was found to bind to prespore vesicles in the posterior cells of *D. discoideum* pseudoplasmodia as well as to the surface of mature spores (Ikeda and Takeuchi, 1971; Takeuchi, 1972). It appears that the antigens present in the prespore vesicles are liberated at the surface during encapsulation. It is assumed that this process liberates mucopolysaccharides and subunits of the spore case which polymerize on the outer surface and form part of the ellipsoid case. During migration, cells in the posterior 70–80% accumulate considerable numbers of these organelles but they seldom appear in the anterior, prestalk cells. This is one of the clearest cases of differentiation of prespore cells in the posterior portions of pseudoplasmodia. The accumulation of prespore vesicles is sensitive to treatment of the cells with actinomycin D and thus seems to depend on RNA synthesis during development (Gregg and Badman, 1970). Prespore vesicles can be seen to arise during late aggregation and do not appear if cells from early aggregates are mechanically dispersed (Gregg, 1971). Small fragments isolated from the anterior of pseudoplasmodia do not accumulate the vesicles unless a period of about 4 hours is allowed for reorganization (Sakai and Takeuchi, 1971). Cells from the posterior prespore portions have been reported to lose the vesicles when incubated as single cells for several hours (Sakai and Takeuchi, 1971), although this has been disputed (Gregg, 1971).

Late in development, several hours after the appearance of prespore vesicles, a heterogeneous group of round membrane-enclosed vesicles with little internal structure have been observed. These vesicles have a distinct appearance due to a peripheral ribosomal layer and have been considered unique to spore differentiation (Gregg and Badman,

---

Fig. 7.5. Stalk wall. The heavy cellulose wall of the stalk is formed from an outer layer (O) 2500 Å thick, and an inner wall (I) which is contiguous with the material between the vacuolized cells. The outer layer is composed of fibers running parallel to the stalk axis which appear as dots in this cross section. The inner layer is amorphous and is presumably laid down by the vacuolizing stalk cells. The surface sheath (S) can be seen to surround the stalk. In places the inner wall is not tightly opposed to the outer layer (H). The ·stalk cells have not fully vacuolized but contain a large central vacuole (V) containing cell debris. The peripheral cytoplasm contains many crystal bodies (X) and some intact mitochondria (M) (TEM).

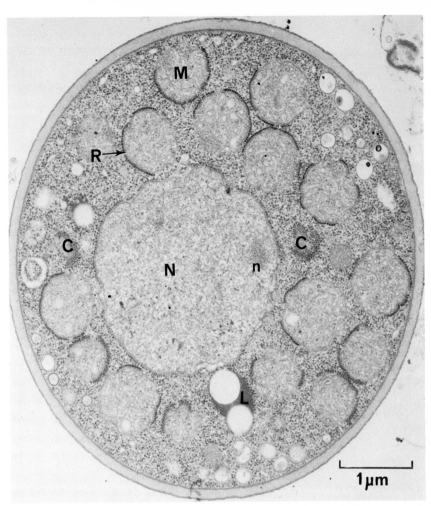

Fig. 7.6. A mature spore. A prominent nucleus (N), with a nucleolar region (n), is evident although slightly crenulate. The nuclear membrane has many attached ribosomes (R). The mitochondria (M) appear rounded and of a uniform size 0.7 μm in diameter. They are also crenulate and decorated with ribosomes which appear as clusters (C) in oblique sections. The mitochondria of spores often contain inclusions (see Fig. 7.7B). Lipid droplets (L) accumulate uniquely in spores (see Fig. 7.7C). Prespore vesicles cannot be seen in mature spores. No homogeneous class of vesicles could be recognized in spores although Gregg and Badman (1970) have described "spore vesicles" they consider specific to spores. The spore wall is shown in more detail in Fig. 7.7A (TEM).

Fig. 7.7. Spore structures. (A) The heavy spore wall is 1000–1500 Å thick and appears to be constructed from an outer amorphous layer 200–300 Å thick, a central fibrous layer 700–900 Å thick, and an inner electron dense layer 100–300 Å thick closely opposed to the plasma membrane. Fixation of thick layers rich in carbohydrate can result in artifacts and the wall may be homogeneously fibrous *in situ.* (B) A mitochondrion of a spore with a prominent inclusion (I). About 5% of the mitochondrial sections prepared from spores show this body which bears no resemblance to the inner matrix of prespore vesicles (see Fig. 7.8), contrary to the assertions of

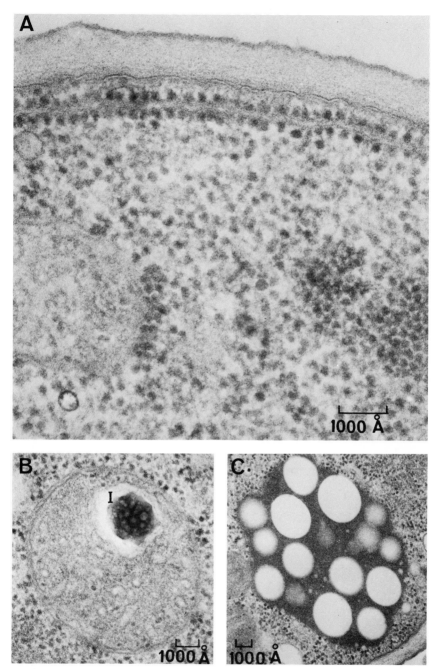

Maeda (1971). (C) A lipid droplet complex. These structures are found exclusively in spores, usually only one per profile. Several electron transparent regions 0.1–0.2 μm in diameter are embedded in electron dense material 0.8–1.2 μm in diameter. The lighter regions may result from elution of more saturated lipids which are less well fixed by osmium tetroxide.

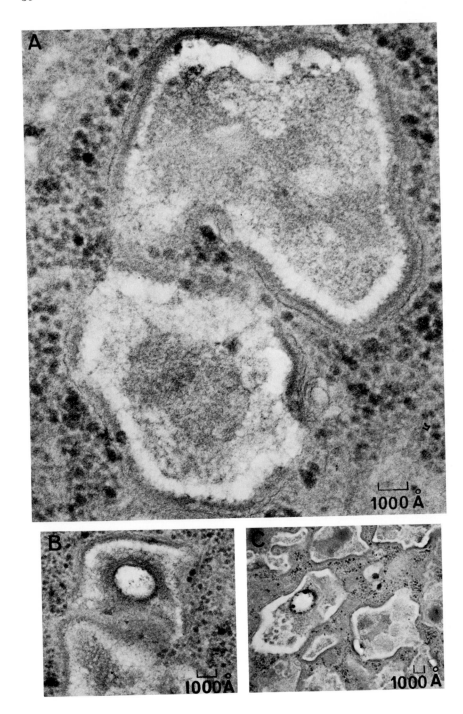

1970). However, similar structures can be found, at a lower frequency, in cells from all stages of development. Moreover, all membrane-bound organelles of spores have a similar peripheral ribosomal layer (Fig. 7.6). It is unlikely that these vesicles represent a unique organelle.

The mechanism which triggers the fusion of prespore vesicles with the plasma membrane is presently unknown. The only pertinent facts available are that spore encapsulation can occur in various mutant strains which fail to form vertical stalks and that it does not occur under any circumstances in isolated cells. Thus, a certain mass of prespore cells appears to be required for encapsulation but there is no strict dependence on the sequence of morphological stages.

Avoidance Reaction

As the anterior tip rises into the air it is oriented such that the stalk is produced at an angle which ensures that the sorus will be as far from surrounding solid objects as possible (Bonner and Dodd, 1962b). On a flat surface the fruiting bodies will be at right angles to the underlying support. Likewise, culmination from a vertical support proceeds horizontally. Clearly the orientation of stalk formation is neither geophobic nor geotropic. When two fruiting bodies culminate within 0.8 mm of each other, the stalks lean away from each other. If the bases actually touch, the angle between the stalks is about 45°. As the distance between the bases increases, the angle decreases. Similar responses can be elicited by the proximity of small glass rods. Orientation of culmination appears to be controlled by negative chemotaxis to a volatile substance given off constantly by the cells themselves. The diffusion gradient of this substance is modified by surrounding barriers so that the highest concentration is in the direction of the barrier. Negative chemotaxis results in avoidance of these obstacles. Likewise, culmination in a "micro wind tunnel" resulted in fruiting bodies pointing up wind (Bonner and Dodd, 1962b). It appears that a gas was blown leeward resulting in a greater concentration of the gas downwind than upwind. The gas can be ab-

---

Fig. 7.8. Prespore vesicles. (A) Vesicles can be observed in cells after the formation of pseudoplasmodia. The vesicles are 0.5–1 μm in diameter and are surrounded by a membrane 60–70 Å thick. A peripheral electron dense layer about 400 Å thick is separated from the central bunch of 50 Å diameter fibers by an electron transparent region of variable size and appearance. This layer may result from condensation of the fibers into a peripheral layer as an artifact of fixation. (B) Some sections of prespore vesicles show a central electron transparent region which may also arise during sample preparation. (C) Prespore vesicles accumulate in large numbers in the posterior cells of pseudoplasmodia shortly before culmination.

sorbed by charcoal but further attempts to characterize it have failed due to the fact that it is active only over very short distances.

Germination Inhibitor

Once encapsulated, the spores are resistant to starvation for periods of months and can withstand dehydration and elevated temperatures (up to 45°C) which kill growing amoebae (Cotter and Raper, 1966, 1968a,b). The spores are also dramatically more resistant to ultrasonic disruption than are growing amoebae (Loomis, 1968). Spores can remain dormant for long periods of time until the sorus is disrupted and the contents diluted. Dormancy is ensured by the accumulation of a germination inhibitor (Ceccarini and Cohen, 1967). The inhibitor can be collected from spores by washing the spores in water and can then be shown to inhibit germination of washed spores when added back to a dilute suspension of spores. It appears to be a heat stable molecule of about 300 molecular weight. Purification of the inhibitor and characterization by mass spectroscopy has suggested that the germination inhibitor of *D. discoideum* is N,N-dimethylguanosine (Bacon *et al.*, 1973). As little as 50 μg/ml of this compound or the native inhibitor have been reported to give 100% inhibition of germination, although this has recently been disputed (Tanaka *et al.*, 1974). The germination inhibitor is not excreted by vegetative or aggregation stage cells but is excreted by the cells during culmination (Fig. 7.9) (Pong and Loomis, unpublished). It will be interesting to determine the biochemical steps in the synthesis of the germination inhibitor since these will most likely be carefully regulated by the overall developmental program.

Germination

When a fruiting body is knocked over by wind, rain, or an encounter with an insect, the spores are dispersed and the germination inhibitor diluted. Germination of the spores then proceeds through a series of stages (Cotter and Raper, 1966). About an hour after activation the spores swell and lose refractility. Shortly thereafter cytoplasmic granules appear and 1 or more contractile vacuoles begin to function. Finally, about 4 hours after activation the spore case is split longitudinally and the amoeba emerges (see Fig. 1.4).

In the absence of a heat shock, germination proceeds only when a source of amino acids is available (Cotter and Raper, 1966). In a medium containing 1% bactopeptone more than 98% of the spores germinate within 7 hours. A mixture of L-tryptophan, L-phenylalanine, and L-methionine

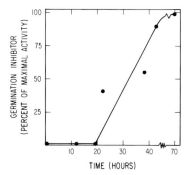

Fig. 7.9. Accumulation of germination inhibitor. *Dictyostelium discoideum* cells were allowed to develop on filter supports at 22°C. At various times samples were collected and the cells suspended at 10^8/ml in SM medium (see Chapter 2). The cells were removed by centrifugation and the supernatant tested for inhibitor activity. Washed spores were suspended at 10^5/ml in SM medium. Within 2 days 60–80% germination was observed. When an equal volume of the supernatant of spores collected at 10^8/ml was added to the medium, no germination occurred. Activity was estimated as the reciprocal of the highest dilution giving 90% inhibition of germination.

allows 90% germination, while L-tryptophan and L-methionine allow only 20% germination in 7 hours. The requirement for these amino acids can be eliminated by heat shocking the spores at 45°C for 30 minutes (Cotter and Raper, 1966, 1968a). After this treatment greater than 90% of the spores germinate within 5 hours in distilled water. Dormancy can be reimposed in heat-shocked spores by addition of the germination inhibitor (Cotter and Raper, 1968b). Germination is sensitive to inhibition of protein synthesis by cycloheximide but is insensitive to the presence of actinomycin D at 250 μg/ml. These results show that *de novo* protein synthesis is necessary after activation for emergence of amoebae and suggest that concomitant RNA synthesis is not required in this process. However, it is not clear that actinomycin D is able to get into spores and inhibit RNA synthesis. The requirement for RNA synthesis after activation is still unclear. Germination appears to be a strictly aerobic process and is inhibited by 5×10^{-4} M dinitrophenol, an uncoupler of oxidative phosphorylation, and by 7.5×10^{-4} M azide which blocks electron transport (Cotter and Raper, 1968b).

Mutational Analysis of Culmination

A mutant strain, KY-3, was isolated some years ago which under normal conditions forms pseudoplasmodia but fails to culminate (Yanagisawa *et al.,* 1967). Pseudoplasmodia of this strain migrate extensively until all of the endogenous reserves are exhausted; fruiting bodies are not formed. During migration, pseudoplasmodia of this strain continuously form stalk which is left behind giving a hairy appearance to the culture. Thus, the major difference between *D. discoideum* and *D. mucoroides,* namely stalked migration, appears to be a mutable characteristic.

Strains have been isolated which form normal migrating pseudoplasmodia but on culmination make little or no stalk. The characteristics of one such strain, KY19, have been described in some detail (Ashworth and Sussman, 1967). The final structure consists of a spherical mound of spores resting on the underlying support.

Strain FR-17 forms normal aggregates, but these fail to transform into migrating pseudoplasmodia (Sonneborn *et al.,* 1963). Instead, the cells differentiate *in situ* into either stalk cells or spore cells with little pattern or polarity. This jumbled morphogenesis results in papillated structures resting as mounds on the support. The derangement of morphogenesis in this strain appears to be a consequence of an alteration in the temporal programming of biochemical differentiations. The strain undergoes all of the biochemical changes observed in the wild type but does so in little more than half the time. Thus, terminal differentiation is complete in strain FR-17 in about 16 hours while fruiting body construction is not normally completed in the wild type until 24 hours after the initiation of development.

The temporal program can also be extended by mutational alteration. A strain, GN-3, has been isolated which passes through the morphological stages in an apparently normal manner but takes 3 days rather than 1 to do so (Loomis, 1970c). Likewise, the biochemical alterations which characterize various stages are dramatically delayed in this strain. This phenotype could result from a general decrease in metabolic processes such as macromolecular synthesis. However, the growth rate of this strain is not significantly different from that of the wild type. It seems more likely that the cells have evolved a mechanism to regulate timing of the sequence of biochemical alterations which lead up to the terminal differentiations and that this mechanism is subject to mutational alteration.

The characterization of several mutant strains has made it clear that the ability of prespore cells to undergo normal encapsulation is not required for fruiting body formation since cells in the sori of these strains remain rounded and do not take on the refractile appearance of spores, yet culmination and stalk formation appear normal (Loomis, 1968; Katz and Sussman, 1972). Defective spores of one of these strains are sensitive to ultrasonic vibration to which wild-type spores are resistant (Loomis, 1968). These strains may carry lesions affecting the synthesis or function of prespore vesicles.

Strains have been isolated in which the spore mass is found at the base rather than on top of the stalk. Careful analysis of the stages in morphogenesis of these strains show that in many cases the spore mass initially rose with the growing stalk but at some point slipped back down the stalk. In other strains the spore mass is never seen to rise off the support. The defect may lie in the construction of the stalk or in the production of adhesive factors by the prespore cells. It has been suggested that a galactose-rich polysaccharide known to accumulate in the prespore mass during culmination serves as an adhesive which holds the sorus in place (M. Sussman and Osborn, 1964). Mutations affecting the synthesis or excretion of this substance would then be expected to result in this phenotype.

Finally, when fruiting body construction is complete, spore cells produce a yellow pigment which permeates the sorus. The pigment has been characterized as a zeta-carotenoid and thus appears to be produced from isoprenoids via the pathway in which squalene is an intermediate (Staples and Gregg, 1967). Mutations which result in white fruiting bodies have been isolated by M. Sussman and R. R. Sussman (1962). Since sorus color is an easily recognized characteristic, the white mutations are convenient to use in genetic crosses (see Chapter 4).

Chapter 8

Macromolecular Aspects of Development

Decrease in Dry Weight and Oxygen Consumption

Development of *D. discoideum* occurs when the exogenous source of nutrients is removed. All subsequent activity and macromolecular synthesis are fueled by endogenous reserves. It is not surprising, there-fore, that the dry weight of *D. discoideum* often goes down by 50% during development (Table 8.1) (Gregg *et al.*, 1954; Gregg and Bron-sweig, 1956a; White and Sussman, 1961; Liddel and Wright, 1961). The initial dry weight of amoebae grown on bacteria varies from 5 to 8 mg per 10^8 cells. The value decreases during development to about 4 mg per 10^8 cells. If the initial values are normalized, the results from numerous experiments suggest that the decrease follows simple first-order kinetics (Fig. 8.1) (White and Sussman, 1961). During the early stages of development, bacteria which have been ingested into food vacuoles

TABLE 8.1

CHANGES IN MACROMOLECULAR COMPLEMENT OF *D. discoideum*[a]

	Amoebae	Fruiting bodies
Dry weight (mg/10^8 cells)	5–8	2.5–4
Protein (mg/10^8 cells)	2–5	1.5–2
RNA (mg/10^8 cells)	1	0.4
Total carbohydrate (mg/10^8 cells)	0.5	0.4
Cellulose (mg/10^8 cells)	0	0.1
Mucopolysaccharide (mg/10^8 cells)	0	0.08
Trehalose (mg/10^8 cells)	0	0.15
DNA (μg/10^8 cells)	36	30

[a] Data for strain NC-4 grown on bacteria from White and Sussman (1961), M. Sussman and Sussman (1969), and Leach and Ashworth (1972).

are degraded and the residue excreted. Thus, the initial drop in dry weight may result from this digestive process.

The rate of oxygen consumption also decreases during development (Gregg, 1950; Liddel and Wright, 1961). On a dry weight basis vegetative cells consume about 0.32 μl/minute oxygen per milligram. The rate decreases to 0.08 μl/minute/mg during culmination (Fig. 8.2). Since total nitrogen decreases somewhat more rapidly than dry weight during development, oxygen consumption per milligram nitrogen increases slightly during the pseudoplasmodial stage and then falls to zero in the fully mature fruiting body (Gregg, 1950). Oxygen consumption is negligible in dormant spores and increases only upon germination.

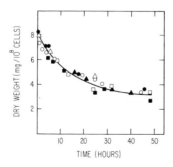

Fig. 8.1. Changes in dry weight during development. Cells developing on agar were collected at various times and dry weight determinations were made. Data from several experiments were normalized to an initial dry weight of 8 mg/10^8 cells. Actual values varied from 5 to 8 mg/10^8 cells (White and Sussman, 1961).

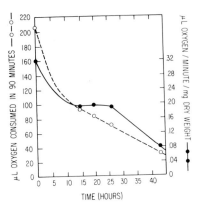

Fig. 8.2. Changes in respiration during development. Cells were allowed to develop in Dixon-Keilin flasks and oxygen uptake measured at various stages. Dry weight determinations were also made during development (Liddel and Wright, 1961).

Protein Content

Protein content of cells grown on bacteria also varies considerably; values from 2 to 5 mg per 10^8 cells have been found (White and Sussman, 1961). These values decrease exponentially during development to a final value of about 2 mg per 10^8 cells in mature fruiting bodies, accounting for about 50% of the dry weight (Fig. 8.3). Thus, the decrease in dry weight results mainly from loss in protein content. Since ammonia and CO_2 are actively given off during development it is likely that most of the protein loss is a consequence of protein degradation and subsequent oxidative metabolism of the amino acids. Several lines of evidence which are discussed in Chapter 11 suggest that the major source of energy during development is derived from amino acid catabolism.

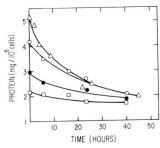

Fig. 8.3. Changes in total protein during development. Each set of points represents a separate experiment (White and Sussman, 1961).

Although total protein content goes down during development, protein synthesis continues throughout development. Attempts have been made to see changes in the spectrum of proteins during development by analyzing the total protein complement. The most sensitive technique used to date, electrophoresis on SDS polyacrylamide gels, separates proteins on the basis of size into several hundred bands but is still of insufficient resolution to clearly discern differences in the pattern of growth stage proteins from that of developmental stage proteins. It is likely that each band contains multiple proteins.

Higher resolution can be achieved by focusing attention on the newly made proteins. Changes in the pattern of synthesis are not obscured by previously formed proteins if the cells are labeled with ^{35}S-methionine for only 2 hours before the proteins are separated on SDS polyacrylamide gels. Autoradiography of the gels after electrophoresis gives an estimate of the amount of the newly made protein in each band.

A schematic representation of the bands which change in intensity during development is given in Fig. 8.4 (Tuchman *et al.*, 1974). There are a large number of discernible bands which are present at each stage but not shown on the figure because no dramatic change could be observed in the gels. Four bands change in a significant manner during development. Bands A and B, which include high molecular weight proteins, are barely discernible over the background in vegetative cells but are intense in the material labeled between 0 and 2 hours after the initiation of development. Synthesis of these proteins decreases gradually during development. Band C can barely be seen either in vegetative

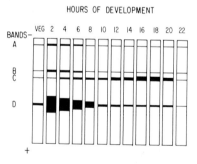

HOURS OF DEVELOPMENT

Fig. 8.4. Schematic representation of changes in the pattern of nascent proteins. Developing cells were labeled with ^{35}S-methionine for 2 hours before being collected. Soluble proteins were dissociated in SDS and separated electrophoretically on SDS acrylamide gels. Autoradiograms of the gels were prepared. A large number of bands was apparent but dramatic changes were seen in only four bands, A, B, C, and D. The width of the bands is proportional to the intensity of the bands on the original autoradiogram (Tuchman *et al.*, 1974).

cells or in the material prepared from cells which had been developing for 2 hours. The intensity gradually increases during subsequent development but decreases rapidly after 18 hours. Band D is the most prominent band on each of the gels. It corresponds to the position at which proteins of 40,000 molecular weight would migrate and appears to consist mostly of *Dictyostelium* actin (Tuchman *et al.*, 1974). The band is present in vegetative cells but increases dramatically in intensity immediately after the initiation of development. The intensity gradually diminishes over the next 6 hours and then drops to the low level seen in vegetative cells during the rest of development.

This approach has been successful in recognizing the developmental changes in actin but has not been able to define changes in other specific proteins during development. An independent approach, that of looking at various enzymes during development, has been used to define the developmental kinetics of a few more proteins and will be discussed in the next chapter.

RNA Content

Growing cells of *D. discoideum* contain about 1 mg RNA per 10^8 cells and this decreases to about 0.4 mg RNA per 10^8 cells during the course of development (White and Sussman, 1961). About 10% of the dry weight of mature fruiting bodies can be accounted for as RNA. Almost all of the RNA in these cells, as in other eukaryotes, is ribosomal RNA (rRNA). Thus, the net decrease in RNA indicates that the ribosomal complement of the cells decreases to about half that of growing cells during the course of development. However, it has been found that *de novo* rRNA synthesis continues long into development, replacing some of those rRNA molecules being degraded (Cocucci and Sussman, 1970). On first consideration it does not seem to make sense for a system which is decreasing its ribosomal complement to continue to synthesize the components. Moreover, even the newly made rRNA molecules are subject to turnover (Cocucci and Sussman, 1970). However, it is possible that the newly made rRNA might differ qualitatively from that synthesized during growth. Accumulation of altered rRNA might modify ribosomal function to adapt it to the specific translational requirements of development.

As in all eukaryotes, there are two major rRNA molecules, the larger associated with the 60 S ribosomal subunit and the smaller with the 40 S subunit. In *D. discoideum* these molecules sediment at 27 S and 17 S, respectively (Ceccarini and Maggio, 1968; Iwabuchi and Ochiai,

1969). The rRNA molecules isolated from cells late in development have the same sedimentation properties and base composition as those isolated from growing cells (Cocucci and Sussman, 1970). Moreover, the species from developed cells cannot be distinguished from those from growing cells by DNA:RNA competition hybridization. Fingerprint analysis of the oligonucleotides formed from rRNA by treatment with specific RNases has indicated no differences in the primary sequences of rRNA synthesized in growing or developed cells (Jacobson and Lodish, personal communication). Thus, by all criteria so far applied there seem to be no differences between the rRNA molecules synthesized at the different stages. We are left with the possibility that *D. discoideum* may not have evolved control mechanisms to stop rRNA synthesis. However, it is still possible that synthesis of rRNA plays some role essential for gene expression during development such as transport of newly made mRNA to the cytoplasm or other processes in macromolecular synthesis of which we are presently unaware.

Polysaccharide Content

Bacterially grown cells contain about 500 μg of carbohydrate per 10^8 cells of which glycogen is one of the major polysaccharides (White and Sussman, 1961). The level of total carbohydrate remains almost constant throughout development. The glycogen content decreases during the first few hours after the initiation of development and then rises severalfold to reach a peak of about 50 μg/10^8 cells just prior to culmination (Fig. 8.5). During fruiting body construction the glycogen content decreases to about half perhaps as a result of its conversion to other polysaccharides. The enzyme which catalyzes the first step in the interconversion and metabolism of glycogen, glycogen phosphorylase, is maximally active during the culmination stage (Firtel and Bonner, 1972b; Jones and Wright, 1970).

Cellulose makes up about 4% of the dry weight of mature fruiting bodies but is undetectable in vegetative cells (Fig. 8.5). It starts to accumulate just prior to culmination as the ring of stalk sheath is laid down just beneath the tip. As more and more stalk sheath is produced during culmination the amount of cellulose increases until finally there are about 100 μg/10^8 cells.

Another polysaccharide, a mucopolysaccharide, accumulates during the culmination stage (Fig. 8.5). It has been characterized as containing N-acetylgalactosamine, galactose, and galacturonic acid and is found associated exclusively with the spore mass (M. Sussman and Osborn,

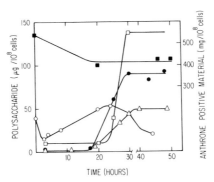

Fig. 8.5. Changes in carbohydrates during development. Amoebae were allowed to develop on agar and were collected at various times for carbohydrate analysis; total anthrone positive material (■———■); trehalose (□———□); glycogen (○———○); cellulose (●———●); mucopolysaccharide (△———△). [Adapted from M. Sussman and Sussman (1969). Patterns of RNA synthesis and of enzyme accumulation and disappearance during cellular slime mould cytodifferentiation. *Symp. Soc. Gen. Microbiol.* **19**, 403–435. Cambridge University Press.]

1964). It appears to be the antigen found initially within prespore vesicles and then found on the external surface of mature spores. It has been proposed that these mucopolysaccharides are responsible for holding the spores together as a viscous mass on the top of the tapered stalk. This substance accounts for about 3% of the dry weight of mature fruiting bodies.

A nonreducing disaccharide, trehalose (glucose-1,1α-D-glucopyranoside), accumulates in spores and appears to serve as an energy reserve used during germination (Ceccarini and Filosa, 1965). It is barely detectable in growing cells and only starts to accumulate at the culmination stage (Fig. 8.5). In mature fruiting bodies the spores contain about 150 μg per 10^8 cells of trehalose which accounts for 6% of the dry weight. Shortly after spores have been activated to germinate, the activity of the enzyme trehalase increases dramatically and appears to lead to the rapid utilization of the stored trehalose (Cotter and Raper, 1970).

Together these polysaccharides make up about 13% of the dry weight of mature fruiting bodies (M. Sussman and Sussman, 1969). Total carbohydrate decreases during development from 500 μg/10^8 cells to 400 μg/10^8 cells of which 300 μg/10^8 cells can be accounted for by glycogen, cellulose, mucopolysaccharide, and trehalose. The remaining 100 μg/10^8 cells can be accounted for by the ribose moieties of the RNA. Essentially all of the major carbohydrate components of differentiated cells of *D. discoideum* have been recognized.

Since the total carbohydrate content of the cells varies little during development, it is not necessary to postulate significant rates of gluconeo-genesis from amino acids. Simple conversion of carbohydrate found in growing cells to the specialized polysaccharides found in the differenti-ated structures could account for the observed developmental kinetics. However, a major source of carbohydrate present in vegetative cells is the ribose moiety of RNA. During development about 200 μg of ribose is liberated by degradation of RNA and the monosaccharide is most likely converted into one of the hexose polysaccharides of the mature fruiting body. Assuming the conversion of pentoses to hexoses proceeds by a biochemical pathway similar to that known to occur in other organ-isms, we would expect considerable metabolic flow from triose phos-phates to hexose phosphates.

DNA Content

The DNA content of *D. discoideum* is one of the lowest of any eukaryotic organism. Cells grown on bacteria contain about 36 μg DNA per 10^8 cells or 0.36 pg per cell, some of which may be of bacterial origin (Ashworth and Watts, 1970; Leach and Ashworth, 1972). This can be compared to 6 pg per cell in humans. In *D. discoideum* the DNA makes up less than 1% of the dry weight of the cells. About a quarter of the DNA in growing cells is of mitochondrial origin but the proportion of this component varies depending on the growth condi-tions of the amoebae.

The total amount of DNA per cell decreases about 15% during the first few hours of development probably as a consequence of degradation of bacterial DNA present in food vacuoles at the time of harvesting the amoebae from growth plates. Leach and Ashworth (1972) have shown that 16% of the DNA of bacterially grown cells is of bacterial origin. There is no change in DNA content throughout the rest of development.

Since *D. discoideum* differentiates in the absence of a food source, no net increase in the total number of cells would be expected. In fact little or no cell division occurs after the initiation of development or during later stages of morphologenesis (Bonner and Frascella, 1952). A few cells in mitosis were observed shortly after the cells were placed in a nonnutrient environment and again during the pseudoplasmodial stage. However, the total number of such cells never exceeded 1% of the population. It is likely that these cells were on the brink of entering mitosis when the amoebae were harvested from growth plates and that they succeeded in entering mitosis several hours later. Recently, it has been shown that the cells are arrested in early G_2 phase during development (Katz and Bourguignon, 1974). When development is initiated in a population of cells in late G_2, the cells proceed through cell division and have a burst of DNA synthesis about 6 hours later. The total cell number doubles within the first few hours but this does not seem to affect the process of aggregation. Cells collected in the S phase do not undergo cell division and yet aggregate with the same kinetics as those collected in middle or late G_2. Thus, although completion of the cell cycle up to early G_2 appears to occur normally, neither DNA synthesis nor cell division is required for development to proceed.

The complete separation of the growth phase from the morphopoietic phase is one of the attractive features of *D. discoideum* as a developmental system. Because of this separation of phases one does not have to consider the effects of differential growth of specialized cells in the analysis of biochemical differentiations in this organism. Moreover, there seems to be no requirement for either DNA turnover or specialized synthesis during development, such as differential gene amplification, since the presence of the drug fluorodeoxyuridine (FUdR) at a concentration which blocks cell growth does not affect development of the cells (Loomis, 1971). Morphogenesis is also insensitive to bromodeoxyuridine (BUdR) at concentrations which inhibit growth and cell division of *D. discoideum* (Loomis, 1971).

Axenically Grown Cells

The macromolecular composition of *D. discoideum* cells grown axenically in HL-5 medium differs considerably from that of bacterially grown cells (Ashworth and Watts, 1970). Although only certain mutant strains can grow axenically in HL-5 medium, the differences do not result from strain variations between the axenic and wild-type cultures but reflect the nutritional conditions (Table 8.2). Strain Ax-2 grown axenically contains almost twice the amount of protein per cell as cells of this strain grown on bacteria. The carbohydrate content is about 5–10 times that of cells grown on bacteria and 94% of this is accounted for by glycogen polymers containing both α-$(1 \rightarrow 4)$ and α-$(1 \rightarrow 6)$ glucosidic linkages (Ashworth and Watts, 1970). The RNA content of axenically grown cells is similar to that of cells grown on bacteria. However, the DNA content of axenically grown cells is less than half of that of cells grown on bacteria perhaps because the majority of cells are found in G_1 phase of the life cycle and so contain only a single copy of the genome (Table 8.2). The DNA content per cell has been reported to decrease during

TABLE 8.2

Macromolecular Composition of Vegetative Cells[a]

Macromolecules	NC-4 on bacteria	Ax-2 on bacteria	Ax-2 on HL-5 (axenic)
Protein (mg/10⁸ cells)	7.8	6.7	11.0
Carbohydrate (mg/10⁸ cells)	0.42	0.46	2.15
RNA (μg of ribose/10⁸ cells)	298	248	328
DNA (μg/10⁸ cells)	36.7	36.0	17.4

[a] From Ashworth and Watts (1970) and Leach and Ashworth (1972). See also R. R. Sussman and Rayner (1971).

development from 17 μg to 13 μg per 10^8 cells (Leach and Ashworth, 1972). The decrease may result from degradation of some of the complement of mitochondrial DNA.

The most striking difference between cells grown axenically and those grown on bacteria is the cellular content of glycogen. Amoebae growing exponentially in HL-5 medium contain about 2.15 mg glycogen per 10^8 cells and this increases up to 5.6 mg per 10^8 cells as the cells enter the stationary phase (Weeks and Ashworth, 1972). These cells develop, nevertheless, in a manner similar to cells grown on bacteria. Thus the initial concentration of glycogen does not seem to affect the developmental program in a significant way.

Axenically grown cells which initiate development with more than 1 mg glycogen per 10^8 cells rapidly reduce their content of cellular glycogen during the aggregation and pseudoplasmodial stages such that mature fruiting bodies contain only about 0.08 mg of glycogen per 10^8 cells (Hames *et al.*, 1972). This is only slightly more than the glycogen content of cells grown on bacteria. Thus cells starting with a 50-fold difference in glycogen are able to regulate their content during development such that the amount of glycogen in mature fruiting bodies is optimal. Removal of excess glycogen appears to result in part from degradation and metabolism of the carbohydrates to CO_2.

Degradation of glycogen during aggregation of axenically grown cells appears to be catalyzed by an amylase rather than glycogen phosphorylase, since the latter enzyme is undetectable in growing cells (Hames *et al.*, 1972; Firtel and Bonner, 1972b) while the former is present at an appreciable level in vegetative amoebae (Jones and Wright, 1970). Even while the degradation of glycogen is proceeding, axenically grown cells start to synthesize other glycogen molecules (Hames *et al.*, 1972). The rate of glycogen synthesis reaches a peak of about 100 μg glycogen per 10^8 cells per hour during the pseudoplasmodial stage and then decreases (Fig. 8.6). The period of maximum synthesis coincides with the accumulation of glycogen in pseudoplasmodia. Unlike glycogen of vegetative cells, glycogen which accumulates specifically during development appears to be insensitive to attack by β-amylase although it is degraded by α-amylase (White and Sussman, 1963a). Therefore, it seems likely that glycogen synthesized during the pseudoplasmodial stage is either qualitatively different from that in growing cells or is localized in a specific subcellular compartment. There may be little or no physiological relationship between these two pools of glycogen.

Although both cells grown on bacteria and axenically grown cells synthesize glycogen at an accelerated rate during the pseudoplasmodial stage, the activity of glycogen synthetase is almost constant throughout development (Wright and Dahlberg, 1967; Hames et al., 1972). Thus, the increased rate of glycogen synthesis appears to result either from activation of the enzyme, an increase in substrate availability, or a modification of subcellular compartments such that substrate and enzyme can interact.

The higher content of protein and glycogen found in axenically grown cells is partly the result of the larger size of these cells (Table 8.2). Moreover, cells collected in the stationary phase of growth in HL-5 medium are considerably denser than those grown on bacteria probably as a result of the high glycogen content (Leach et al., 1973). Nevertheless, cell size decreases in axenically grown cells during development such that the resulting spores are actually smaller than those of cells grown on bacteria (Table 8.3). The 10- to 20-fold decrease in cell volume

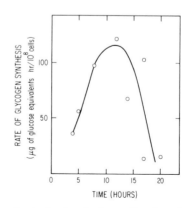

Fig. 8.6. Glycogen synthesis in developing cells. Amoebae of strain Ax-2 were allowed to develop on filter supports. The cells initially contained 2.6 mg of glycogen/10^8 cells (Hames et al., 1972).

TABLE 8.3
CELL SIZE[a]

Strain	Growth conditions	Amoebae (μm^3)	Spores (μm^3)
NC-4	Bacteria	695	112
Ax-2	Bacteria	695	64
Ax-2	Axenically in HL-5 medium	1165	48

[a] Calculated from data of Ashworth and Watts (1970) and Leach and Ashworth (1972).

during spore development necessitates removal of a large amount of the cell mass originally present. Thus, glycogen removal in axenically grown cells may be essential for adaptation of cell size to that appropriate for sporulation. In cells grown on bacteria the restriction of cell size is not as demanding and may occur as a natural consequence of utilization of endogenous reserves in carrying out the specific steps of development.

The breakdown of protein and RNA and subsequent oxidation of the components appear to be the major source of energy during development. Moreover, subunits for macromolecular biosynthesis must be derived from turnover of vegetative macromolecules. As is discussed below, specific RNA molecules and specific proteins have been shown to accumulate during discrete stages of development.

RNA Metabolism

Dictyostelium discoideum is a convenient eukaryotic organism in which to investigate the steps of RNA synthesis, transport, and function. Multiple RNA polymerases have been isolated from both growing and developing cells, the nuclear heterogeneous RNA has been isolated and analyzed, and changes in mRNA complement at various stages have been measured.

The RNA polymerases of *D. discoideum* resemble those isolated from a variety of other eukaryotes including sea urchins, *Xenopus*, and calf thymus (Pong and Loomis, 1973a). There are two major enzymes both of about 420,000 molecular weight. Polymerase I is found predominantly in nucleoli and is insensitive to the drug, α-amanitin. It makes up about 20–30% of the total extractable activity from both vegetative and developing cells. Because of its subcellular localization and α-amanitin insensitivity it is thought to preferentially direct the synthesis of ribosomal RNA. The remaining two-thirds of the extractable activity is polymerase II which is sensitive to α-amanitin and appears to direct the synthesis of mRNA. The 2 polymerases can also be distinguished by differences in ionic sensitivity: polymerase I acts optimally in the presence of 0.075 M KCl, while polymerase II acts optimally in the presence of 0.06 M (NH_4)$_2SO_4$. The enzymes can be separated by chromotography on DEAE-cellulose. Polymerase II has been purified to greater than 90% homogeneity and was found to consist of subunits of 190,000, 170,000, 150,000, 28,000, 21,000, and 15,000 daltons in molar ratios of one (Pong and Loomis, 1973a). The subunit structure is the same in polymerase II isolated from either vegetative or developing cells. One can calculate that there are about 32,000 molecules of polymerase II per cell during growth and this decreases to about 8000 molecules per cell during development. Thus, during development the cells reduce their capacity to catalyze the synthesis of RNA by about 75%.

The study of RNA polymerases in *D. discoideum* was extended to search for possible factors which could stimulate activity of specific genes, such as the CRP protein appears to do on catabolic genes of bacteria. Crude nuclear and cytoplasmic fractions were tested on the enzymes, using several DNA templates, but showed no significant stimulation whether the fractions were derived from growing or developing cells (Noble and Loomis, unpublished). Likewise, fractions from G-100 gel filtration of the proteins of the nucleus failed to stimulate the activity of either polymerase I or II in a significant manner. Thus, this approach has not yet been able to shed light on the mechanism by which the pattern of transcription might change during development. Perhaps if the analysis is extended to the study of RNA synthesized from specific DNA sequences, it may be possible to recognize activator or repressor molecules.

Structure of Nuclear and Cytoplasmic RNA

The majority of the RNA transcribed in exponentially growing cells is a 36 S precursor of ribosomal RNA (Fig. 8.7) (Cocucci and Sussman, 1970; Mizukami and Iwabuchi, 1972; Firtel and Lodish, 1973). Nuclei contain in addition to the 36 S precursor, processed 27 S and 17 S ribosomal RNA as well as a 19 S precursor of the smaller rRNA. There is also some labeled material sedimenting in the region from 4 to 17 S. The ribosomal precursors are found only in the nucleus where they appear to be processed before moving to the cytoplasm. In the cytoplasm

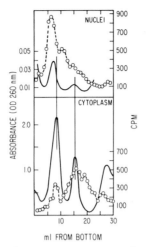

ml FROM BOTTOM

Fig. 8.7. Size distribution of RNA in growing amoebae. Cells of strain Ax-1 were grown in HL-5 medium to a titer of about 10^7 cells/ml. During the final 15 minutes the cells were labeled with ^{3}H-uridine. Nuclei and soluble cytoplasm were separated and treated with 1% SDS before being layered on a 13–23% sucrose gradient containing 0.1 *M* NaCl, 0.01 *M* Tris buffer pH 7.4 and 0.5% SDS. After centrifugation for 17 hours at 22,500 rpm in an SW-25 rotor, samples were collected from the bottom and analyzed for OD_{260} (solid line) and radioactivity (O———O). Data are given in amounts per 10^8 cells (Cocucci and Sussman, 1970).

the newly made RNA is mostly 27 S and 17 S ribosomal RNA. However, with a very short pulse, about 25% of the labeled material is seen to be heterogeneous in size with a peak at about 14 S (Fig. 8.8). Greater than 92% of the labeled material in the cytoplasm is associated with polysomes (Firtel and Lodish, in press).

The heterogeneous polysomal RNA contains a sequence of about 100 adenosine residues at the 3′ end (Firtel, Jacobson *et al.*, 1972). As in other eukaryotes, the poly A sequence does not appear to be part of the primary transcription product but is attached to the 3′ end of completed mRNA in the nucleus before entering the cytoplasm. The poly A associated RNA can be separated from bulk RNA by its high affinity to poly U held either on filters or attached to Sepharose. The poly A associated RNA molecules sediment in the region between 5 and 28 S (Fig. 8.8). Hybridization kinetics of this material have shown that greater than 85% hybridizes to unique sequences present as single copies in the genome.

The heterogeneous nuclear RNA (HnRNA) also contains poly A sequences (Firtel and Lodish, 1973). The poly A associated RNA of the nucleus appears to be about 20% larger than cytoplasmic mRNA although

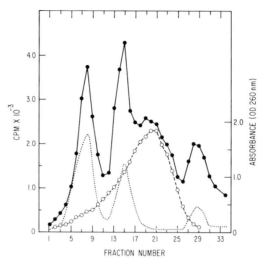

Fig. 8.8. Poly A associated RNA of *D. discoideum*. Exponentially growing cells were labeled for 18 minutes with ^{3}H-uracil. The cells were broken with detergent and the extracted RNA was sedimented through a 15–30% sucrose gradient containing 0.1% SDS by centrifugation for 21 hours at 26,500 rpm in a Spinco SW 27.1. Samples were collected from the bottom and aliquots counted immediately (●——●) or collected and washed on a filter to which poly U was fixed (○——○) (Firtel and Lodish, 1973).

there is a wide distribution of sizes. The average size of HnRNA of *D. discoideum* is about 500,000 daltons. The very large RNA molecules sedimenting at greater than 35 S which are found in the nuclei of mammalian cells are not found in *D. discoideum*. Since most of the large HnRNA of mammalian cells is degraded in the nucleus and never appears in the cytoplasm, large portions of these giant molecules are not translated. In contrast, in *D. discoideum* more than two-thirds of the HnRNA synthesized in the nucleus appears in the cytoplasm within 15 minutes after the addition of actinomycin D. It is clear that most of the poly A associated RNA in the nucleus is a direct precursor of mRNA.

The kinetics of hybridization of nuclear poly A associated RNA has shown that about 70% is transcribed from DNA which is present as single copies in the genome (Firtel and Lodish, 1973). Moreover, about 25% of the sequences present in nuclear heterogeneous RNA hybridize to repetitive sequences present in several hundred copies per genome. Mild alkali treatment resulting in single breaks in the HnRNA molecules separated the poly A segments from much of the sequences transcribed from repetitive DNA. These results show that most RNA molecules contain a sequence on the 5′ end of the molecule which is transcribed from repetitive DNA. Since only 12% of the cytoplasmic mRNA is complementary to repetitive DNA, much of the 5′ sequence transcribed on repetitive DNA must be removed before the RNA enters the cytoplasm. The repetitive sequences must be at or near the 5′ end of the molecule rather than internal in the molecule since HnRNA decreases only about 20% in size before entering the cytoplasm. This reduction can be accounted for if about 300 bases transcribed from a common genetic element are trimmed from the 5′ end prior to transport to the cytoplasm.

At the 3′ end of mRNA molecules there is a run of about 25 adenosine molecules (Jacobson *et al.*, 1974a,b). There appears to be an equal number of these oligo A sequences and the poly A sequences in mRNA which indicates that the oligo A is present in all of the poly A associated RNA. Unlike the poly A segment, the oligo A sequence is transcribed from the DNA template. *Dictyostelium discoideum* DNA contains about 15,000 sequences of oligo dT which are 25 bases long. In contrast there are no detectable sequences of poly dT which are 100 bases long that could conceivably code for poly A segments of mRNA. The oligo dT segments, 25 bases long, are dispersed throughout the genome and appear to be transcribed at the 3′ end of each mRNA molecule. The RNA product synthesized *in vitro* by isolated *D. discoideum* nuclei also contains oligo A segments at the 3′ end (Jacobson *et al.*, 1974a,b). The steps in HnRNA synthesis and processing are summarized in Fig. 8.9.

There do not appear to be any differences in the steps of mRNA

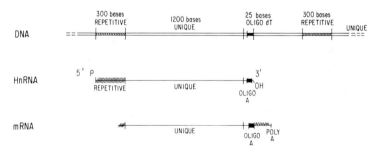

Fig. 8.9. Proposed structure of a gene and its transcriptional products. A gene of 1200 bases is preceded by a repetitive region of about 300 bases. The repetitive sequence is transcribed into the initial product, HnRNA, but these bases are subsequently cleaved off during formation of mRNA. A similar region precedes another unique gene. At the other end of the gene there is a run of 25 thymine bases which is transcribed into oligo A. It is suggested that most or all genes contain an oligo T sequence. A few bases separate the oligo A sequence from the poly A sequence added posttranscriptionally (Firtel and Lodish, 1973).

synthesis between growing and developing cells. Moreover, the average size of mRNA is the same at both stages (13–14 S) (Firtel and Lodish, 1973). As was shown for the mRNA of growing cells, the poly A segment in developing cells is about 100 nucleotides long. Developing cells synthesize both 27 S and 17 S rRNA but these enter the cytoplasm much more slowly than in growing cells. The proportion of newly synthesized cytoplasmic RNA which is associated with poly A is about 15% in growing cells and the proportion increases to 60% during development (Firtel and Lodish, 1973). Thus there appears to be considerably less ribosomal RNA synthesized during development. This presents a well-defined case of differential transcriptional control of a specific gene, that of coding for rRNA.

Transcriptional Changes

The pattern of protein synthesis at any given stage in development will depend in part on the complement of mRNA present at that stage. Since different proteins are synthesized at different stages in *D. discoideum,* we would expect different portions of the genome to be represented as RNA transcripts during the different stages. The only alternative would be a mechanism of selective translation in which control mechanisms functioned to single out certain mRNA species from a complex mixture of mRNA species and allowed translation to proceed from those molecules. There is no evidence at present which indicates such posttranscriptional control in *D. discoideum.*

Since mRNA molecules are known to hybridize to unique sequences present only once in the genome, it has been possible to determine the proportion of the genome which is expressed at different stages of development (Firtel, 1972) (Table 8.4). Single copy DNA was separated from repetitive sequences in DNA and sheared to pieces of about 400 nucleotides. The strands of DNA were separated by thermal denaturation and then allowed to reanneal for a limited time. Repetitive sequences reformed DNA:DNA duplexes during this time due to the higher initial concentration of these sequences. Single copy sequences could then be separated from duplexes on the basis of binding to hydroxyapatite (Firtel, 1972). Labeled single copy DNA was incubated in the presence of a vast excess of RNA isolated from various stages. RNA molecules present in only one copy per cell were in sufficient concentration to form RNA:DNA duplexes under the conditions of the experiment. The proportion of DNA which formed RNA:DNA duplexes gave an estimate of the percent of the genome which had transcripts present at that stage.

TABLE 8.4
STAGE-SPECIFIC TRANSCRIPTION

Source of RNA	% of Genome present as RNA transcripts[a]	Estimated number of genes expressed[b]
Growth	30.2	8,600
Aggregation	27.4	7,800
Pseudoplasmodium	34.2	10,000
Culmination	35.6	11,000
All stages	56.0	16,000

[a] Data from Firtel (1972). Calculated on the assumption that only one DNA strand directs transcription.

[b] Estimated for genes of 1000 nucleotides each.

RNA from vegetative amoebae hybridized to 15.1% of the single copy DNA. Assuming only one strand of DNA is transcribed, 30.2% of the unique proportion of the genome is transcribed during the vegetative stage. From the reannealing kinetics of the unique portion of the genome it can be calculated that there are about 2.86×10^7 nucleotide pairs in single copy DNA (Firtel, 1972). This represents enough information for 28,600 genes of 1000 nucleotides each. About 30% of this information is expressed during growth and can be accounted for by transcription of about 8600 genes.

The proportion of the genome expressed decreases slightly during aggregation and then increases during the latter stages of development (Table 8.4). However, the increase in number of single copy transcripts late in development does not represent only reexpression of genes functioning during growth but appears to result from expression of genes functioning specifically during development. While RNA from the growth stage binds 15.1% of the single copy DNA, and RNA from the culmination stage binds 17.8% of the single copy DNA, RNA molecules from the two stages incubated together with single copy DNA bind 24.7% of the DNA. Clearly new genes are expressed during development.

DNA sequences expressed during growth could be isolated as hybrids formed from single copy DNA and RNA of growing amoebae. When these were freed of RNA by alkali hydrolysis, the percent which could bind to RNA isolated from the culmination stage was determined. Only about half of the genes expressed during growth still had transcripts present at culmination (Firtel, 1972). Similar experiments showed that slightly more than 60% of the transcripts present in culmination were not present during the growth stage.

These estimates are subject to a variety of sources of error; however, they are internally consistent and do not appear to be subject to large systematic errors. The estimates of the proportion of the genome represented at various stages show that about 5600 genes are expressed in each of the stages and may direct the synthesis of proteins with "housekeeping" function.

Another 3000 genes are expressed in growing cells but their transcripts gradually disappear during development. These genes may direct the synthesis of proteins involved in growth functions such as mitosis, cell division, and DNA synthesis.

Finally, there are about 7000 genes which are expressed exclusively during development. The accumulation of RNA transcripts from these genes is gradual and occurs throughout the developmental stages. It is likely that these genes direct the synthesis of the stage-specific proteins necessary for aggregation, pseudoplasmodium formation, stalk formation, and spore encapsulation.

Altogether almost 60% of the unique portion of the genome is expressed at one stage or another. This is a considerably higher proportion than has been found in mammalian systems and may indicate that there are selection pressures on *D. discoideum* which restrict the genome to close to the minimum size compatible with its physiological complexity. Unlike metazoan organisms, *D. discoideum* passes much of its life cycle as rapidly dividing single cells. Under these conditions excess DNA may be of selective disadvantage. It is not known if the remaining 40% of the unique sequences is transcriptionally silent or is expressed under conditions which have not yet been analyzed, such as during germination or encystment.

Chapter 9

Enzymatic Aspects of Development

Many of the proteins which accumulate during discrete stages are enzymes which catalyze specific reactions. Alterations in the enzymatic complement of the cells may modify physiological processes and lead to terminal differentiations. Although only a small percentage of the total enzymatic complement of *D. discoideum* has been analyzed at any stage of development, observation on a few dozen enzymes has demonstrated the variety of mechanisms for enzymatic adaption which have evolved in this system.

Enzymes of the Growth Phase

Some of the enzymes present in vegetative cells disappear shortly after the initiation of development (Table 9.1). The enzyme trehalase catalyzes the hydrolysis of trehalose and appears to play a significant role in mobilizing it as an energy source during germination (Ceccarini, 1966). However, during development trehalose is not used as an energy source but is accumulated for utilization later during germination of the spores. Consistent with this pattern of trehalose storage, trehalase

TABLE 9.1
SOME ENZYMES OF *D. discoideum*

Enzyme	Reference
Growth[a]	
1. Trehalase	Ceccarini, 1967
2. Threonine deaminase-1	Pong and Loomis, 1973b
3. Acid DNase	M. Sussman and Sussman, 1969
4. Alkaline DNase	M. Sussman and Sussman, 1969
5. β-Glucosidase-1	Coston and Loomis, 1969
Growth and Development[b]	
1. Adenyl cyclase	Rossomando and Sussman, 1972
2. Pyrophosphatase	Gezelius and Wright, 1965
3. Lactic dehydrogenase	Firtel and Brackenbury, 1972
4. Aspartate transaminase	Firtel and Brackenbury, 1972
5. Glutamic transaminase	Firtel and Brackenbury, 1972
6. Homoserine dehydrogenase	Firtel and Brackenbury, 1972
7. Phospholipase A	Ferber *et al.*, 1970
8. Lysophospholipase	Ferber *et al.*, 1970
9. Protease	M. Sussman and Sussman, 1969
10. Acid amylase	Jones and Wright, 1970
11. Neutral amylase	Jones and Wright, 1970
12. α-Glucosidase	Pollock and Loomis, unpublished
13. Glucokinase	Baumann, 1969
14. Glucose-6-phosphate dehydrogenase	Wright, 1960
15. 6-Phosphogluconate dehydrogenase	Edmundson and Ashworth, 1972
16. Embden-Meyerhof enzymes	Cleland and Coe, 1968
17. Fructose-1,6-diphosphatase	Baumann and Wright, 1969
18. Glycogen synthetase	Hames *et al.*, 1972; Wright and Dahlberg, 1967
Development[c]	
Partially characterized enzymes:	
1. Acid phosphatases	Wiener and Ashworth, 1970
2. GDP-mannose transferase	Bauer *et al.*, 1971
3. Cellulase	Rosness, 1968
4. cAMP phosphodiesterase	Riedel *et al.*, 1973
Stage-specific enzymes:	
1. Leucine aminopeptidase	Firtel and Brackenbury, 1972
2. Alanine transaminase	Firtel and Brackenbury, 1972
3. N-Acetylglucosaminidase	Loomis, 1969a
4. α-Mannosidase	Loomis, 1970a
5. Trehalose-phosphate synthetase	Roth and Sussman, 1968
6. Threonine deaminase-2	Pong and Loomis, 1973b
7. Tyrosine transaminase	Pong and Loomis, 1971
8. UDPG pyrophosphorylase	Ashworth and Sussman, 1967
9. UDPgalactose polysaccharide transferase	M. Sussman and Osborn, 1964
10. UDPG epimerase	Telser and Sussman, 1971
11. Glycogen phosphorylase	Firtel and Bonner, 1972b; Jones and Wright, 1970
12. Alkaline phosphatase-2	Loomis, 1969c
13. β-Glucosidase-2	Coston and Loomis, 1969

[a] Enzymes decrease during development.
[b] Activity changes less than 3 fold.
[c] Activity increases during development.

is excreted from the cells during aggregation and is not resynthesized until late in culmination or during germination (Cotter and Raper, 1970; Killick and Wright, 1972b).

Amino acid metabolism changes quite dramatically during the life cycle of *D. discoideum*. During growth there is a demand for biosynthesis while during development amino acids are a major energy source. This change is reflected in the rapid loss of the enzyme threonine deaminase-1 during development (Pong and Loomis, 1973b). From a variety of observations it appears that this enzyme plays a biosynthetic role catalyzing the first step in the conversion of L-threonine to L-isoleucine. The decreased demand for net synthesis of isoleucine is paralleled by decreased synthetic activity. The biosynthetic enzyme is not only inactivated but also replaced later during development by a catabolic enzyme which degrades threonine. This enzyme, threonine deaminase-2, is discussed below.

Two DNase enzymes have been observed in cells growing on bacteria, one acting in an acidic environment, and one in an alkaline environment (M. Sussman and Sussman, 1969). These enzymes probably attack the DNA of ingested bacteria. Following the initiation of development when the bacterial food source is removed there is little need for such enzymes and so it is not surprising that they disappear.

β-Glucosidase-1 is a lysosomal enzyme which may be involved in bacterial degradation (Coston and Loomis, 1969). It appears to play no significant role during development since mutant strains specifically lacking β-glucosidase-1 have been isolated and found to undergo morphogenesis in a completely normal manner (Dimond and Loomis, in press). It is present in growing amoebae but is not synthesized following the initiation of development (Coston and Loomis, 1969). The specific activity of β-glucosidase-1 increases initially as the total protein content of the cells decreases. About 4 hours after the initiation of development the specific activity drops precipitously such that very little activity remains at the end of development. If either RNA or protein synthesis is blocked during development, the decrease in β-glucosidase-1 activity is partially inhibited suggesting that processes involving macromolecular synthesis are necessary for removal of β-glucosidase-1. The actual mechanisms of inactivation of this enzyme are not clear but appear to occur following excretion of the enzyme.

Enzymes of Growth and Development

Many enzymes appear to function during both growth and development. These enzymes have been found to be present at all stages of the life cycle of *D. discoideum* and to remain at essentially the same specific activity throughout development (Table 9.1).

Rossomando and Sussman (1972) have shown that adenyl cyclase, the enzyme responsible for the synthesis of cAMP, is present in growing cells and does not increase during aggregation although the cells secrete cAMP at a high rate only during aggregation (see Chapter 6). These authors concluded that if adenyl cyclase plays a significant role in the control of cAMP production, then modulation of its activity rather than its synthesis must be coupled to the aggregation process. The enzyme appears to be localized in the plasma membrane (Rossomando, 1974).

Pyrophosphatase is present at high activity in growing cells and the specific activity remains unchanged during development (Gezelius and Wright, 1965). The enzyme hydrolyzes pyrophosphate liberated in the synthesis of cAMP and many other compounds. It can be thought of as a scavenger enzyme recycling phosphate generated in biosynthetic reactions. Lactic dehydrogenase probably plays a similar scavenging role and is present during both growth and development (Firtel and Brackenbury, 1972). Three enzymes of amino acid metabolism have been found to remain constant even though amino acid catabolism becomes the major energy source shortly after the initiation of development. Aspartate transaminase, glutamate dehydrogenase, and homoserine dehydrogenase all remain at essentially the same specific activity throughout development (Firtel and Brackenbury, 1972). Perhaps the initial activity is sufficient for the metabolic demands which occur during development.

Vegetative cells have an extremely high content of the phospholipid degrading enzymes, phospholipase A and lysophospholipase (Ferber *et al.*, 1970). The activity remains high during development. These observations suggest that a high level of membrane degradation or turnover occurs at all stages in this system. Likewise, an acid protease has been found to remain at the same specific activity throughout development (M. Sussman and Sussman, 1969). Protein turnover occurs at all stages in this organism at about 8% per hour (Wright and Anderson, 1960a,b; Franke and Sussman, 1973).

Jones and Wright (1970) have described two amylases, an acid, and a neutral enzyme which are present at about the same specific activity at all stages of development. The acid amylase may be identical to an α-glucosidase assayed with *p*-nitrophenyl α-glucoside as substrate (Pollock and Loomis, unpublished). The latter enzyme has a pH optimum of 5 and is present throughout development. It appears to be subject to balanced turnover during most of development since, when its synthesis is blocked with cycloheximide, the specific activity drops fairly rapidly. Extensive investigations have failed to discover any differences in the physical or enzymatic properties of the enzyme from growing or developing cells and thus it is likely that α-glucosidase is identical at all stages. This seems to be a case in which synthesis of a specific enzyme continues in a constitutive manner thoughout the life cycle. During growth it may aid in the breakdown of bacterial glycogen while during development it appears responsible for removal of stored glycogen.

Glucose liberated from glycogen can be readily phosphorylated by the enzyme, glucokinase, which shows little difference in activity between growing amoebae and culminating fruiting bodies (Baumann, 1969). The enzyme is highly specific for glucose. Further metabolism of the glucose 6-phosphate formed can occur via the Embden-Meyerhof pathway or via 6-phosphogluconate to ribose phosphate formation. The enzymes glucose-6-phosphate dehydrogenase and 6-phosphogluconate dehydrogenase are present at all stages (Wright, 1960; Edmundson and Ashworth, 1972).

Cleland and Coe (1968) did a survey of all of the enzymes of the Embden-Meyerof pathway in vegetative, aggregating, and culminating cells. Although there were changes in certain of the enzymes, their specific activities were found to change less than 3-fold. The estimated overall activity for the pathway appears to remain constant throughout development.

Enzymes of the Embden-Meyerhof pathway catalyze reactions involved in gluconeogenesis as well as carbohydrate degradation. However, gluconeogenesis requires the action of fructose-1,6-diphosphatase. This enzyme is present at all stages at a low but constant level (Cleland and Coe, 1968; Baumann and Wright, 1969). The enzyme has a high affinity for fructose-1,6-diphosphate ($K_m = 6 \times 10^{-5}$ M) and so may be able to handle the flow of metabolites leading to hexoses without a requirement for a large pool of the substrate.

Gluconeogenesis leads to accumulation of glycogen in the pseudoplasmodial stage. However, the increase in the rate of glycogen synthesis does not appear to result from an increase in the biosynthetic enzyme since the specific activity of glycogen synthetase changes little during development (Wright and Dahlberg, 1967; Hames *et al.*, 1972). It appears that factors other than enzyme availability control glycogen synthesis in *D. discoideum*. It has been considered that the availability of the substrate, UDP glucose, or a specific effector may control the overall metabolic pattern.

Enzymes of Development

From a developmental point of view the enzymes which accumulate during specific stages are probably the most interesting (Table 9.1).

Analysis of the kinetics and mode of synthesis of these enzymes is facilitated by the restricted period of accumulation. One is tempted to relate the increase in a specific enzyme to newly acquired functions of the differentiating cells. In some cases the relation seems to be justified while in others the connection appears to be indirect and complex. The stage-specific enzymes can also be used as molecular markers which indicate the biochemical processes going on in a carefully regulated temporal sequence during development. If these enzymes are to be used to monitor specific gene expression it is imperative to know that each activity results from a single enzyme and not from a family of isozymes, otherwise one might be monitoring a complex multigenic pattern rather than the expression of a single gene. Purification of the enzymes has often indicated whether a given activity results from the combined action of several isozymes. However, more direct evidence can be provided by isolation and analysis of mutations in the structural genes. To date, these approaches have been applied to some but not all of the stage-specific enzymes.

Analysis of the acid phosphatase activity in *D. discoideum* has indicated that there are at least 5 isozymes (Solomon *et al.*, 1964). These appear to accumulate sequentially during development and require concomitant protein synthesis (Wiener and Ashworth, 1970; Loomis, unpublished). However, the difficulties inherent in the study of such a complex of isozymes have precluded further study.

GDP-mannose transferase accumulates to a specific activity about 4-fold higher than the basal level shortly after aggregation (Bauer *et al.*, 1971). The enzyme is associated with a subcellular particle and requires a polymer acceptor. These properties have raised technical barriers to further analysis and we know little more than these few facts about this enzyme.

The specific activity of cellulase rises from an undetectable level to a peak during culmination (Rosness, 1968). However, when extracts of pseudoplasmodial cells were mixed with extracts of culminating cells, activity in the latter cells was inhibited. Thus, it appears the low level of activity measured in cells before culmination is due to the presence of an inhibitor and so we have no way of knowing how much actual cellulase is present in the early stages of development. The inhibitor is a heat labile protein which appears to act by binding to the enzyme. Since the observed enzymatic activity results from the interactions of 2 or more proteins, cellulase does not provide a useful marker for monitoring gene expression during development and has not been studied further. The enzyme may actually function during germination when the amoebae must emerge from the cellulose-containing spore cases.

The kinetics of accumulation of cAMP phosphodiesterase have been determined both in cells growing in suspension on bacteria and in cells developing in suspension cultures (Riedel *et al.*, 1973; Malkinson and Ashworth, 1973). The activity of the extracellular enzyme is low during exponential growth of the cells. As the cells reach stationary stage the enzyme activity increases more than 20-fold. However, when the food source is exhausted the activity drops precipitously. The decrease in activity is not due to degradation of the enzyme but to the accumulation of a specific protein inhibitor which inactivates the enzyme. This inhibitor accumulates during the first few hours following starvation of the cells and thus appears to be a stage-specific protein itself. The kinetics of accumulation of cAMP phosphodiesterase can be most clearly observed in mutant strains lacking the specific inhibitor. In these strains the enzyme continues to accumulate for several hours after starvation of the cells and is stable thereafter. Several mutant strains which fail to aggregate do not accumulate cAMP phosphodiesterase. Cells developing on filter supports accumulate the enzyme after about 3 hours of development (Malkinson and Ashworth, 1973). The specific activity increases about 20-fold by 9 hours and then decreases to about half the peak-specific activity during the following 6 hours. About half the recoverable activity is found extracellularly throughout development indicating that the cells secrete this enzyme. The role that this enzyme may play in aggregation of *D. discoideum* is discussed in Chapter 5.

Marker Enzymes of Development

The remaining 13 stage-specific enzymes have all been studied in considerable detail and appear to represent single gene products. All of the enzymes have been shown to require concomitant protein synthesis and a prior period of RNA synthesis. The developmental kinetics of these enzymes are given in Fig. 9.1.

Enzyme Synthesis

Detailed studies on these enzymes have shown that the assays are strictly linear with the amount of enzyme in extracts taken throughout development and have elucidated the exact kinetics of accumulation under conditions of normal and abnormal morphogenesis. The accumulation has been shown in each case to be blocked when protein synthesis is inhibited by the drug cycloheximide. An example of this analysis showing the effect of addition of cycloheximide on accumulation of α-

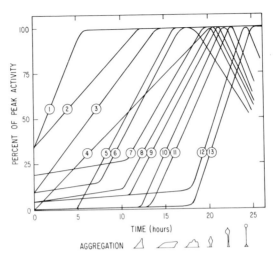

Fig. 9.1. Stage-specific enzymes of *D. discoideum*. Development was initiated by depositing bacterially grown amoebae on filter supports. Samples were taken and the specific activities of the following enzymes were determined:

Code	Enzyme	Abbreviation
1	Alanine transaminase	AT
2	Leucine aminopeptidase	LAP
3	N-Acetylglucosaminidase	NAG
4	α-Mannosidase	MAN
5	Trehalose-phosphate synthetase	TPS
6	Threonine deaminase-2	TD
7	Tyrosine transaminase	TT
8	UDPG pyrophosphorylase	PP'ase
9	UDPgal polysaccharide transferase	Trans
10	UDPG epimerase	EPI
11	Glycogen phosphorylase	GP
12	Alkaline phosphatase	Alk P
13	β-Glucosidase-2	BG-2

See Table 9.1 for appropriate references.

mannosidase during development is given in Fig. 9.2. Similar results have been found when the other stage-specific enzymes have been analyzed, which suggest that accumulation of each requires concomitant protein synthesis and that simple activation is not responsible for the increase in activity.

There are at least two ways in which preferential accumulation of a specific protein can occur: either an increase in the rate of synthesis

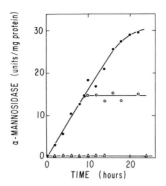

Fig. 9.2. Effect of cycloheximide on the accumulation of α-mannosidase. Cells of *D. discoideum* NC-4 developing on filter supports were treated with 400 μg/ml cycloheximide at 0 (△) or 9 hours (○) after the initiation of development. Control cells developed in the absence of the drug (●) (Loomis, 1970a).

or a decrease in the rate of degradation can result in accumulation. No matter which mechanism is functioning, *de novo* protein synthesis is required. However, if a preferential decrease in degradation is responsible for the accumulation, the enzyme must turn over very rapidly before the period of accumulation. When *de novo* protein synthesis is blocked with cycloheximide the activities of the stage-specific enzymes do not decrease under most conditions. Hence, there is no evidence for rapid turnover of the enzymes and it would appear that accumulation results from an increased rate of synthesis. It has been suggested that cycloheximide itself not only blocks protein synthesis but also affects protein degradation directly. However, there is no evidence for such an effect in *D. discoideum* or in any other cellular system.

The mechanism of accumulation can be directly determined by labeling cells for very short periods throughout development and determining the amount of synthesis and degradation of specific enzymes. This approach requires that one totally purify the enzyme with good yield which often involves considerable difficulties. So far the approach has been applied only to the enzyme UDPG pyrophosphorylase. This enzyme is present at a low but constant level during the aggregation stage and then accumulates up to 10-fold during the pseudoplasmodial stage (Ashworth and Sussman, 1967). Cells labeled with radioactive amino acids during the aggregation stage incorporated very little radioactivity into UDPG pyrophosphorylase while cells labeled during the pseudoplasmodial stage incorporated considerable label into the enzyme (Fig. 9.3) (Franke and Sussman, 1973). UDPG pyrophosphorylase was purified by precipitation from crude extracts with antiserum prepared against

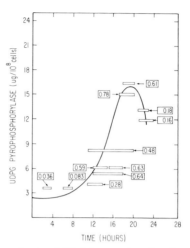

Fig. 9.3. Differential synthesis of UDPG pyrophosphorylase. The kinetics of accumulation are indicated by the solid line. Cells were labeled with ^{14}C-amino acids during the periods indicated by the boxes. The amount of label incorporated into purified UDPG pyrophosphorylase is given next to each box as the fraction of label incorporated into total proteins (Franke and Sussman, 1973).

purified enzyme. The immune precipitate was then dissociated in detergent and the subunits separated by sodium dodecyl sulfate polyacrylamide gel electrophoresis. A prominent band occurred in the position at which authentic subunits of the enzyme migrated electrophoretically. Radioactivity under this band was taken as a measure of the amount of newly synthesized enzyme. The results of a series of such experiments suggested that during aggregation, both synthesis and degradation of the enzyme occurred at a rate of about 0.1 μg/10^8 cells/hour. During the pseudoplasmodial stage the degradation rate remained unchanged while the rate of synthesis increased to 1.5 μg/10^8 cells/hour (Franke and Sussman, 1973). Although conflicting results have appeared (Gustafson and Wright, 1972, 1973a; Gustafson *et al.*, 1973), it appears that accumulation of UDPG pyrophosphorylase results from a sharp increase in the rate of synthesis during pseudoplasmodial stage with little or no change in the rate of degradation throughout development.

While similar experiments have yet to be performed on the other stage-specific enzymes, it is likely that they also accumulate due to an increased rate of synthesis during specific stages. An increase in the rate of synthesis of an enzyme requires either the accumulation of mRNA for that enzyme or more efficient utilization of preexisting mRNA. It is not presently possible to directly determine the concentration of a specific mRNA in *D. discoideum* and so the experiments have had to be somewhat indirect. The hybridization experiments discussed in the previous chapter show that the global content of mRNA changes during development, but do not indicate whether mRNA molecules for specific enzymes have changed. If the increase in enzyme synthesis resulted from an increased rate of translation on preformed stable mRNA, then inhibition of RNA synthesis would not be expected to affect enzyme accumulation. This possibility was tested by treating the cells with 125 μg/ml actinomycin D at various stages in development and observing subsequent enzyme accumulation. The drug blocks total RNA synthesis more than 95% within 15 minutes both *in vivo* and *in vitro* at this concentration and has no immediate effect on polysome size or protein synthesis (Sussman *et al.*, 1967; Pong and Loomis, 1973a; Firtel and Lodish, in press). Lower levels of the drug do not affect mRNA synthesis but block rRNA synthesis. Actinomycin at a concentration of 125 μg/ml does not affect the half-life of poly A associated mRNA at any stage of development (Firtel and Lodish, in press).

When RNA synthesis is blocked by the addition of actinomycin D, the accumulation of most of the stage-specific enzymes continues uninterrupted for about 2 hours and then stops abruptly suggesting that continued transcription is required for accumulation of these enzymes and that there is a limited pool of mRNA which can function for a short period in the presence of the drug. A typical response is seen in the accumulation of α-mannosidase (Fig. 9.4). The decay of enzyme-forming

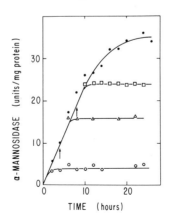

Fig. 9.4. Effect of actinomycin D on the accumulation of α-mannosidase. Cells of *D. discoideum* NC-4 developing on filter supports were treated with 125 μg/ml actinomycin D at 0 (●), 4 (△), and 8 hours (□) after the initiation of development. The arrows indicate the time of addition of the drug. Control cells developed in the absence of the drug (○) (Loomis, 1970a).

capacity in the presence of actinomycin D does not follow first-order kinetics as it does in bacteria where it is thought that the rate of decay of mRNA depends upon the initial concentration of the specific mRNA. In *D. discoideum* the kinetics suggest that there may be two pools of mRNA with greatly different stabilities. Over a 2-hour period there is entry of mRNA from the relatively stable pool to the unstable pool. One possibility is that there is a relatively stable pool of mRNA in the nucleus which slowly enters in unstable cytoplasmic pool over a 2-hour period.

The period between the time of addition of actinomycin D and the time at which further increase in enzyme-specific activity stopped was found to be about 2 hours for most of the stage-specific enzymes. However, for UDPG pyrophosphorylase, UDP galactose polysaccharide transferase, and alkaline phosphatase, the periods were found to be 7 hours, 5 hours, and 14 hours, respectively (Roth et al., 1968; Loomis, 1969c). This was interpreted as indicating posttranscriptional control of the mRNA for these enzymes. However, recent studies have shown that actinomycin D does not block all poly A associated RNA (mRNA) synthesis even at the doses utilized (125 μg/ml) (Firtel et al., 1973). Genes which are less sensitive to the effects of the drug may therefore continue to synthesize mRNA in the presence of actinomycin. When daunamycin is added together with actinomycin D, mRNA synthesis is inhibited more than 98% at all stages (Firtel et al., 1973). Under these conditions the accumulation of each of the stage-specific enzymes is blocked when the drugs are added 2 hours before the period of accumulation of the specific enzyme. When the drugs are added at later times, accumulation continues for about 2 hours and then stops abruptly. If the total enzyme accumulated is plotted at the time of inhibition of RNA synthesis, we can see the relation of enzyme forming capacity to enzyme synthesis (Fig. 9.5). The most direct interpretation of these results is that transcription of mRNA for each of the stage-specific enzymes occurs about 2 hours prior to the period of accumulation and then proceeds in parallel with the synthesis of the enzymes.

In any case, it is clear that the accumulation of each of the stage-specific enzymes requires a period of RNA synthesis shortly before the period of enzyme synthesis. This observation is consistent with a model of control at the transcriptional level such that the differential rate of mRNA synthesis increases during a specific stage. However, these observations do not rule out the alternative possibility that the differential rate of transcription does not itself change during development but that the rate of degradation of specific mRNA molecules decreases during a particular stage. This mechanism requires considerable turnover of

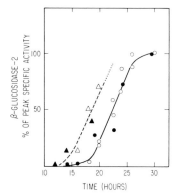

Fig. 9.5. Capacity to accumulate β-glucosidase-2 in the presence of actinomycin D. At various times after the initiation of development cells were treated with 125 μg/ml actinomycin D. Samples were taken during the subsequent 24 hours to determine the peak specific activity of β-glucosidase-2 estimated electrophoretically (▲) or as the net increase in β-glucosidase specific activity after 16 hours (△). The peak specific activity of the treated cells is plotted at the time of addition of the drug. Untreated controls were assayed for β-glucosidase-2 estimated electrophoretically (●) or as the net increase in specific activity after 16 hours (○). These values are plotted at the time the sample was collected (Coston and Loomis, 1969).

mRNA except at discrete stages. Direct measurement of nuclear poly A associated RNA shows that very little turnover occurs and that more than 80% leaves the nucleus and becomes associated with polysomes (Firtel and Lodish, 1973). Differential stability of mRNA cannot account for the altered rates of enzyme synthesis in this system. The increase in rate of synthesis of specific enzymes most likely results from transcriptional controls which stimulate the synthesis of specific mRNA species during discrete stages.

Enzyme Degradation

As can be seen in Fig. 9.1, the specific activities of several of the stage-specific enzymes decrease after reaching a peak. If protein synthesis is blocked by cycloheximide prior to attainment of peak activity, previously accumulated activity is stable, but if cycloheximide is added after the attainment of peak activity, inactivation proceeds at the control rate. The same results are found if actinomycin D is added during the period of accumulation. It seems that specific developmental events are necessary for the inactivation of some enzymes during discrete periods. These processes require RNA and protein synthesis shortly before the cessation of enzyme synthesis. Thus, although there is no evidence for rapid turn-over of the stage-specific enzymes during the period of accumulation, there is evidence for specific inactivation of some of the enzymes at the end of the synthetic period. Sussman and Lovgren (1965) have pointed out the significance of such systems to differentiating organisms whose growth is insufficient to reduce the specific activity by dilution with other proteins. The only recourse is to inactivate the enzymes whose continued functioning might not be compatible with the physiology of subsequent stages in differentiation.

The mechanisms of inactivation are known for only a few of the stage-specific enzymes. In the case of UDPgalactose polysaccharide transferase, an enzyme which accumulates during culmination, the decrease in specific activity late in culmination appears to result from preferential excretion of the enzyme followed by inactivation on the outside (Sussman and Lovgran, 1965). Since the polysaccharide product of this enzyme has been found in the prespore vesicles, it is likely that the enzyme is also localized in these organelles. The prespore vesicles appear to fuse with the plasma membrane and eject their contents during encapsulation of the spores (see Chapter 7). UDPgalactose polysaccharide transferase may also be ejected in this process.

The partial loss of enzyme activity seen in late stages of development for several of the stage-specific enzymes appears to be a consequence of loss of activity in the vacuolized stalk cells. The rapid swelling and death of these cells are likely to inactivate enzymes present in the cells. When stalk differentiation is blocked, as in mutant strain KY19, the activity of UDPG pyrophosphorylase is not inactivated. It will be interesting to determine the mechanisms which inactivate other stage-specific enzymes that accumulate at other periods.

Stage-Specific Enzymes

Since the well-characterized, stage-specific enzymes form convenient biochemical markers during the development of *D. discoideum*, the characteristics of the enzymes and the kinetics of accumulation in bacterially grown cells induced to develop synchronously on filter supports will be discussed in some detail. Enzyme activity for the stage-specific enzymes is expressed in units defined as that amount of enzyme which catalyzes the formation of 1 nmole of product per minute under defined conditions. For convenience, they will be categorized as (1) enzymes of amino acid metabolism, (2) hydrolases, and (3) enzymes of carbohydrate metabolism.

Enzymes of Amino Acid Metabolism (Fig. 9.6)

Leucine Aminopeptidase (EC 3.4.1.2). Immediately after the removal of the exogenous food source, the specific activity of leucine aminopeptidase increases from a level of 8 units/mg protein to reach a peak of about 30 units/mg protein 16 hours later during the pseudoplasmodial stage. The specific activity remains at this high level throughout the rest of development (Firtel and Brackenbury, 1972). The enzyme is assayed with the chromogenic substrate L-leucine p-nitranilide at pH 7.2 at 30°C. The enzyme sediments at 5.6 S. Assuming a globular configuration the molecular weight is about 85,000. It migrates as a single peak on polyacrylamide electrophoresis. From a consideration of the substrate specificity of the enzyme, it may play a role in the degradation of peptides to single amino acids during development.

Alanine Transaminase (EC 2.6.1.2). This enzyme is present in cells growing exponentially on bacteria at a specific activity of 200 units/mg protein and accumulates to a level of 500 units/mg protein during the first 5 hours of development (Firtel and Brackenbury, 1972). The specific activity remains constant thereafter. Extracts are prepared in 15% glycerol containing 0.2 mM dithiothreitol and assayed at 28°C with L-alanine, α-ketoglutarate, lactic dehydrogenase, and NADH at pH 7. Oxidation of NADH by the pyruvate generated from transamination of alanine is measured spectrophotometrically. By sedimentation analysis, alanine

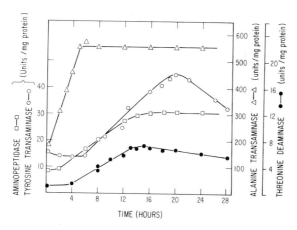

Fig. 9.6. Enzymes of amino acid metabolism. Cells of *D. discoideum* NC-4 were allowed to develop on filter supports. Samples were taken for the assay of alanine transaminase (△——△); aminopeptidase (□——□); threonine deaminase-2 (●——●); and tyrosine transaminase (○——○) (Firtel and Brackenbury, 1972; Pong and Loomis, 1971; Pong and Loomis, 1973b).

transaminase appears to be a single protein of about 6 S (100,000 daltons). It is probably involved in amino acid metabolism during development.

Threonine Deaminase-2 (*EC 4.2.1.16*). Some difficulty is encountered in assaying for threonine deaminase-2 due to the presence of a biosynthetic enzyme, threonine deaminase-1, in vegetative cells (Pong and Loomis, 1973b). However, the biosynthetic enzyme can be preferentially inactivated by freezing and thawing. The biosynthetic enzyme is sensitive to feedback inhibition by L-isoleucine and L-leucine, while threonine deaminase-2 is insensitive to these amino acids. By these characteristics, one can distinguish the different isozymes during development. The biosynthetic enzyme disappears during the first 8 hours of development while the catabolic enzyme, threonine deaminase-2, accumulates from an initial specific activity of 1 unit/mg protein to a peak of 7 units/mg protein at 16 hours during the pseudoplasmodial stage. The specific activity decreases slightly during the remaining hours of development. Samples are collected in the presence of pyridoxal phosphate and frozen. The enzyme is assayed with L-threonine at pH 8.5 at 25°C and the ketone product recognized as the 2,4 dinitrophenyl hydrozone derivative. The enzyme sediments as a single band of activity at 9.5 S (M.W. = 120,000). Threonine deaminase-2 can also catalyze the catabolism of L-serine and is probably involved in metabolism of both of threonine and serine.

Tyrosine Transaminase (*EC 2.6.1.5*). This enzyme accumulates from a level of 15 units/mg protein to a peak of 45 units/mg protein between 8 and 20 hours of development (Pong and Loomis, 1971). The specific activity decreases about 25% late in development. The enzyme is assayed with pyridoxal phosphate, α-ketoglutarate and L-tyrosine at pH 8.5 at 32°C. The product is measured spectrophotometrically at 331 nm. The enzyme sediments as a symmetrical peak at 5 S (M.W. = 80,000). Extensive comparisons of physical and enzymatic properties of the enzyme present in vegetative cells to those present in developing cells have failed to detect any differences. Synthesis of this enzyme occurs during growth, stops during aggregation, and recommences during the pseudoplasmodial stage.

Hydrolases (Fig. 9.7)

Acetylgucosaminidase (*EC 3.2.1.30*). Initiation of development results in a rapid increase of acetylglucosaminidase from a basal level of 20 units/mg protein to a level of about 220 units/mg protein within 12 hours (Loomis, 1969a). The activity is stable throughout the rest of

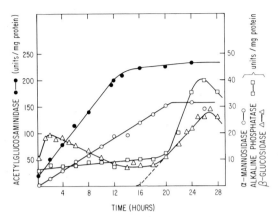

Fig. 9.7. Hydrolases. Cells of *D. discoideum* NC-4 were allowed to develop on filter supports. Samples were taken for assay of acetylglucosaminidase (●——●); alkaline phosphatase (□——□); α-mannosidase (○——○); and β-glucosidase (△——△) (Loomis, 1969a,c, 1970a; Coston and Loomis, 1969).

development. The enzyme is assayed with the chromogenic substrate, *p*-nitrophenyl *N*-acetylglucosaminide at pH 5 at 35°C. The enzyme has an apparent molecular weight of 168,000 with subunits of 65,000 and 55,000 (Every and Ashworth, 1973). A single gene accounts for 99% of the maximal activity (Dimond *et al.*, 1973). The remaining 1% is a minor isozyme with distinct properties (Dimond and Loomis, 1974). Acetylglucosaminidase is localized in lysosomes and excreted during both growth and during migration. It is able to hydrolyze both *N*-acetylglucosamine and *N*-acetylgalactosamine moieties in internal glycosidic linkages. It appears to play an essential role during migration since mutations in the *N*-acetylglucosaminidase gene result in small, slowly migrating pseudoplasmodia (Dimond *et al.*, 1973).

α-Mannosidase (*EC 3.2.1.24*). This enzyme also accumulates soon after the initiation of development but does not reach maximal specific activity until after 20 hours of development (Loomis, 1970a). The specific activity is less than 0.1 units/mg protein at the start of development but reaches about 35 units/mg protein at 20 hours. It is stable thereafter. The enzyme is assayed with the chromogenic substrate *p*-nitrophenyl α-mannoside at pH 5 at 35°C. The enzyme has an apparent molecular weight of 280,000 (Every and Ashworth, 1973). Two mutants have been isolated in which less than 0.1% of the normal activity accumulates during the aggregation or pseudoplasmodial stages (Free and Loomis, 1975). During culmination a minor isozyme of α-mannosidase accumulates in these strains to a peak specific activity of about 6 units/mg protein and thus makes up about 17% of the activity in wild-type fruiting bodies. The minor isozyme has a 10-fold lower K_m for the substrate than the major isozyme and has a higher pH optimum. Like other acid hydrolases the major isozyme of α-mannosidase is localized in lysosomes and excreted during migration.

Alkaline Phosphatase (EC 3.1.3.1). This enzyme accumulates late in development between 18 and 26 hours after removal of the bacterial food source (Loomis, 1969c). A distinct isozyme is present at a level of 5 units/mg protein in growing cells and remains at about the same specific activity throughout development. At 18 hours the specific activity is 10 units/mg protein and this increases to a peak of 40 units/mg protein during culmination. The specific activity decreases by a factor of 2 during the subsequent 10 hours. Both isozymes appear to be membrane-bound but can be released by treatment with nonionic detergents (McLeod and Loomis, in press). The solubilized isozymes differ in sedimentation properties; the early isozyme sediments at 9 S while the late isozyme sediments at about 12 S. The isozymes can also be separated electrophoretically (Krivanek, 1956). The isozymes are assayed with *p*-nitrophenyl phosphate at pH 8.5 at 25°C in the presence of $MgCl_2$. Alkaline phosphatase-2, the second isozyme, is able to hydrolyze several phosphorylated compounds including 5′-AMP (Gezelius and Wright, 1965). Accumulation of alkaline phosphatase-2 appears to require the multicellular state found in culminating fruiting bodies since cells dissociated at 18 hours accumulate much less enzyme. Since this stage-specific enzyme accumulates so late in development, it is likely that it plays a role in germination rather than culmination.

β-Glucosidase-2 (EC 3.2.1.21). There are two easily separated isozymes of β-glucosidase in *D. discoideum* (Coston and Loomis, 1969). The first isozyme disappears during the aggregation stage while the second isozyme accumulates between 18 and 30 hours of development. There is a slight loss of activity during the next 10 hours. The specific activity at 18 hours is 8 units/mg protein while at 30 hours the specific activity is 30 units/mg protein. When the isozymes are electrophoretically separated it can be seen that β-glucosidase-2 increases more than 10-fold during culmination. β-Glucosidase-2 sediments as a sharp peak of activity at 10 S and has an apparent molecular weight of 220,000. The enzyme catalyzes transglycosylation as well as hydrolysis of β-glucosides. It does not appear to be essential for normal culmination and may play a role in germination.

Although the isozymes differ in sedimentation properties and substrate affinity as well as in electrophoretic mobility, they appear to depend on a common genetic element, since 4 strains which were selected for loss of β-glucosidase-1 were found to have also lost the ability to accumulate β-glucosidase-2 (Dimond and Loomis, in press). Other stage-specific enzymes accumulated normally in these strains. It is not yet clear whether the mutated gene codes for a common subunit or is a regulatory element.

Enzymes of Carbohydrate Metabolism (Fig. 9.8)

Trehalose-phosphate Synthetase (*EC 2.3.1.15*). This enzyme catalyzes the reaction:

$$\text{Glucose 6-phosphate} + \text{UDPG} \rightarrowtail \text{trehalose 6-phosphate} + \text{UDP}$$

The specific activity is very low in growing cells (less than 1 unit/mg protein) and remains low during the beginning of aggregation. At about 6 hours the specific activity begins to increase and continues to increase for 10 hours when a peak specific activity of 10 units/mg protein is reached (Roth and Sussman, 1968, Newell *et al.*, 1972). The activity

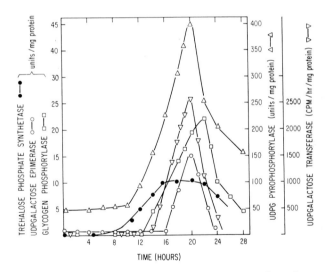

Fig. 9.8. Enzymes of carbohydrate metabolism. Cells of *D. discoideum* NC-4 were allowed to develop on filter supports. Samples were taken for assay of trehalose-phosphate synthetase (●——●); UDPG pyrophosphorylase (△——△); UDPgalactose polysaccharide transferase (▽——▽); UDPgalactose epimerase (○——○); and glycogen phosphorylase (□——□) (Roth and Sussman, 1968; Ashworth and Sussman, 1967; Sussman and Osborn, 1964; Telser and Sussman, 1971; Firtel and Bonner, 1972b).

is seen to drop during later stages of development but it has been reported that the decrease is due to inactivation following extraction and that this can be prevented by the presence of 50 mM trehalose in the extract (Killick and Wright, 1972b). The enzyme is measured at pH 7.0 at 37°C by determining the amount of UDP produced in a coupled reaction with phosphoenol pyruvate and pyruvate kinase. The enzyme appears to be cold sensitive and is inactivated by prolonged storage at 2°C (Killick and Wright, 1972b). It has been reported that the specific activity in early aggregating cells can be increased dramatically by partial purification of the enzyme (Killick and Wright, 1972a). When crude extracts are treated with 60% $(NH_4)_2SO_4$ and the precipitate collected and resuspended, the activity is found to be increased 50-fold. From these observations it has been suggested that trehalose-phosphate synthetase is present at all stages of development of D. discoideum at about the same concentration but that it is "masked" by association with a protein inhibitor early in development. Ammonium sulfate precipitation apparently removes the inhibitor. This curious phenomenon has not been further explored. If this picture is correct, it would imply that exponentially growing cells synthesize both trehalose-phosphate synthetase and its inhibitor throughout the growth stage although the enzyme would be inactive. This possibility is difficult to reconcile with the observations that protein synthesis is required during development for the increase in specific activity and that little increase occurs following the inhibition of RNA synthesis (Roth et al., 1968).

UDPG Pyrophosphorylase (EC 2.7.7.9). More attention has been focused on this enzyme than on any other in D. discoideum since it catalyzes the synthesis of an intermediate, UDPG, which appears to be critical to the biosynthesis of many polysaccharides found in differentiated cells. The activity is low in vegetative cells and increases about 8-fold between 10 and 20 hours of development (Ashworth and Sussman, 1967). The specific activity increases from 45 to 400 units/mg protein when the amount of glucose 1-phosphate formed from UDPG is assayed. Optimal conditions were found to be pH 7.6 in Tricine buffer at 35°C with 4 mM $MgCl_2$ and 2 mM pyrophosphate. The enzyme has been purified to homogeneity and found to have an apparent molecular weight of 390,000 (Franke and Sussman, 1971). The subunit molecular weight is about 55,000. The following K_m values were obtained with the purified enzyme: glucose 1-phosphate, 2.6×10^{-4} M; UTP, 1.1×10^{-4} M; UDP-glucose, 1.7×10^{-4} M; pyrophosphate, 4.4×10^{-4} M (Franke and Sussman, 1971). Binding of UDPG by the enzyme is sensitive to competitive inhibition by UTP. Moreover, pyrophosphate is a noncompetitive inhibitor. The enzyme makes up less than 0.5% of the total protein at peak specific activity. As discussed previously, the increase in specific activity of UDPG pyrophosphorylase appears to result from differential synthesis of the enzyme late in development when polysaccharide synthesis is placing a heavy demand on UDPG.

UDPgalactose Polysaccharide Transferase. From a historical point of view it is interesting that this was the first enzyme shown to be stage specific in *D. discoideum.* It catalyzes the transfer of galactose from UDPgalactose to a polysaccharide containing galactosamine, galacturonic acid, and galactose (Sussman and Osborn, 1964). The enzyme is undetectable in growing cells and accumulates dramatically between 14 and 22 hours of development. It is then excreted from the cells and is inactivated in the extracellular space. The assay depends on the addition of a polysaccharide acceptor purified from culminating cells. The reaction rate is measured by following the transfer of radioactive galactose from UDPgalactose to the polysaccharide at pH 7.5 at 25°C. The rate depends on the particular polysaccharide preparation used and so cannot be expressed in absolute units. The enzyme is undoubtedly necessary for synthesis of the galactose polymer found in prespore vesicles which is later extruded. It is likely that the enzyme is also localized in prespore vesicles. The enzyme accumulates in posterior prespore cells of pseudoplasmodia but is absent in prestalk cells (Newell *et al.,* 1969). This is the only enzyme studied to date which has been shown to be subject to spatial as well as temporal control in *D. discoideum.*

UDPgalactose Epimerase (EC 5.1.3.2). This enzyme is present at a specific activity of 1 unit/mg protein in growing cells but decreases to less than 0.1 unit/mg protein in pseudoplasmodia. Between 16 and 20 hours of development epimerase increases rapidly to a specific activity of 15 units/mg protein (Telser and Sussman, 1971). The enzyme is then rapidly excreted and inactivated such that by 22 hours no measurable activity remains. The activity is assayed by coupling the production of UDPG from UDPgalactose to the enzyme UDPG dehydrogenase and measuring the coupled reduction of NAD^+ spectrophotometrically. The reaction is carried out at pH 8.5 at 37°C. Epimerase is probably responsible for the formation of UDPgalactose which is incorporated into polysaccharide by the action of UDPgalactose transferase in the prespore vesicle. The epimerase accumulates during a very restricted period somewhat after the transferase appears and may catalyze the limiting reaction in mucopolysaccharide synthesis. The enzyme may also be localized in prespore vesicles.

Glycogen Phosphorylase (*EC 2.4.1.1*). The glycogen content of *D. discoideum* decreases late in development when cellulose and mucopolysaccharide accumulate. Therefore, it seems logical that glycogen phosphorylase might be developmentally controlled. Glycogen phosphorylase is low or undetectable in growing cells and accumulates between 12 and 24 hours to a peak specific activity of 22 units/mg protein (Jones and Wright, 1970; Firtel and Bonner, 1972b). The specific activity decreases in mature fruiting bodies until no activity remains 36 hours after the initiation of development. The enzyme is assayed with soluble starch at pH 6.8 at 35°C. Glucose 1-phosphate formed in the reaction is measured by a linked enzyme system with phosphoglucomutase, glucose-6-phosphate dehydrogenase and 6-phosphogluconate dehydrogenase. The reduction of $NADP^+$ is measured spectrophotometrically. Alternatively the reverse reaction can be measured by monitoring the incorporation of radioactive glucose 1-phosphate into glycogen. The activity in the reverse direction is less than a tenth that in the forward direction. The K_m for the glucose 1-phosphate is 1.2 mM and for phosphate is 3.4 mM. UDPG is a competitive inhibitor for both glycogen and inorganic phosphate in the forward reaction. The dissociation constant of the enzyme–UDPG complex (K_i) is 2.9×10^{-4} M. At 5.4×10^{-4} M UDPG, the K_p (effective Michaelis constant in the presence of inhibitor) for phosphate is 9.8 mM. At 1 mM UDPG the K_p for phosphate is 17 mM. The interaction of these effectors is discussed in a subsequent chapter.

Mutational Analysis of Stage-Specific Enzymes

The sequential appearance of stage-specific enzymes during restricted portions of the developmental program suggests that a mechanism for temporal control functions in *Dictyostelium*. Analysis of the developmental kinetics of the enzymes in two temporally deranged mutant strains supports this view. One strain, FR-17, develops almost twice as fast as the wild type while another strain, GN-3, develops more than twice as slowly as the wild type (Sonneborn *et al.*, 1963; Loomis, 1970c). An analysis of 5 stage-specific enzymes in strain FR-17 has shown that each one accumulates precociously (Fig. 9.9). The temporal derangement is more marked in the enzymes of late development which suggests that an acceleration of the temporal controls occurs during development of strain FR-17. Although the enzymes appear precociously, they appear in the normal order. This observation indicates that each enzyme is subject to an underlying temporal control. Alternatively it is possible that cascade inductions are taking place in which the appearance of each enzyme depends on the accumulation of prior enzymes.

Seven enzymes have been followed in strain GN-3 and all have been found to be delayed (Fig. 9.9). Cells of this strain pass through each

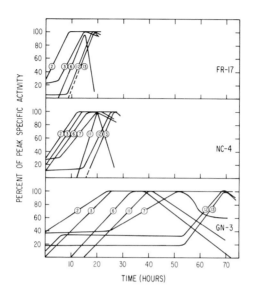

Fig. 9.9. Temporal control in mutant strains. The developmental kinetics of amino-peptidase (2), acetylglucosaminidase (3), threonine deaminase (6), tyrosine trans-aminase (7), glycogen phosphorylase (11), alkaline phosphatase (12), and β-glucosi-dase-2 (13) were determined in a fast mutant, FR-17, and a slow mutant, GN-3. The kinetics of these enzymes in wild-type strain NC-4 are given for comparison (Firtel and Brackenbury, 1972; Loomis, 1969c, 1970c; Pong and Loomis, 1971; Firtel and Bonner, 1972a,b; Coston and Loomis, 1969).

of the morphological stages observed in the wild type but take almost 3 days to do so rather than a single day. The early enzymes, aminopep-tidase and N-acetylglucosaminidase, accumulate during the initial pro-longed period of aggregation and do not reach peak specific activity until after 24 hours. Threonine deaminase is also proportionally delayed. Although the order of biochemical events is almost normal in strain GN-3 and most of the enzymes appear in conjunction with the appropri-ate morphological stages, glycogen phosphorylase and tyrosine trans-aminase appear in reverse order (Fig. 9.9). Normally glycogen phos-phorylase does not begin to accumulate until 6 hours after tyrosine transaminase begins to accumulate. However, in strain GN-3, glycogen phosphorylase reaches peak specific activity almost 10 hours before tyro-

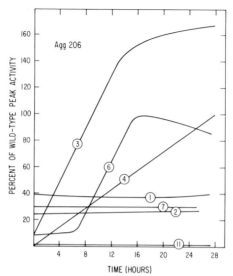

Fig. 9.10. Biochemical differentiations in an aggregateless strain. The developmental kinetics of alanine transaminase (1), aminopeptidase (2), acetylglucosaminidase (3), α-mannosidase (4), threonine deaminase (6), tyrosine transaminase (7), and glycogen phosphorylase (11) were determined in aggregateless strain Agg 206. (See Table 9.1 for appropriate references; Loomis, 1974.)

sine transaminase has accumulated fully. This observation suggests that temporal controls for these enzymes are distinct and modified to different extents by the mutation in strain GN-3. However, these results were collected in separate experiments in two different laboratories and it is possible that different culture conditions affected the rate of development of strain GN-3.

A fairly large number of morphological mutants has been analyzed for the appearance of stage-specific enzymes so as to delineate some of the steps which can be genetically modified. For instance strain Agg 206 which fails to aggregate or undergo any of the visual stages of morphogenesis still accumulates N-acetylglucosaminidase, α-mannosidase, and threonine deaminase-2 almost normally (Fig. 9.10). Later enzymes such as tyrosine transaminase and glycogen phosphorylase do not accumulate as a pleiotropic response to the lesion in aggregation. More surprising is the observation that neither aminopeptidase nor alanine transaminase accumulates in this strain (Fig. 9.10). In the wild-type strain these are the first enzymes to accumulate. Yet the mutation in strain Agg 206 blocks their accumulation without affecting the later enzymes, acetylglucosaminidase, α-mannosidase, or threonine deaminase.

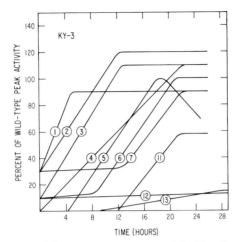

Fig. 9.11. Biochemical differentiations in Strain KY3. The developmental kinetics of alanine transaminase (1), aminopeptidase (2), acetylglucosaminidase (3), α-mannosidase (4), trehalose-phosphate synthetase (5), threonine deaminase (6), tyrosine transaminase (7), glycogen phosphorylase (11), alkaline phosphatase (12), and β-glucosidase-2 (13) were determined in the culmination defective strain KY-3. (See Table 9.1 for appropriate references; Loomis, 1974).

The most likely explanation is that accumulation of acetylglucosaminidase, α-mannosidase, and threonine deaminase is a delayed response to an event which occurs prior to the signal for accumulation of aminopeptidase and alanine transaminase. Other aggregateless strains do not accumulate acetylglucosaminidase or threonine deaminase.

A strain which forms pseudoplasmodia but fails to culminate, strain KY-3, accumulates all of the stage-specific enzymes analyzed except for those of culmination, alkaline phosphatase and β-glucosidase-2 (Fig. 9.11). Another mutant strain, min 2, which culminates normally but fails to encapsulate spores, nevertheless, accumulates alkaline phosphatase and β-glucosidase-2.

The pattern of accumulation can be ordered with a unique polarity suggesting a linear pattern to the pleiotropic effects of the morphological mutations. Figure 9.12 indicates that the pattern of biochemical differentiations can be subdivided into at least 5 stages on the basis of which enzymes accumulate in the various strains.

Three further developmental stages can be defined by the morphological behavior of these mutant strains (Fig. 9.13). Strains TS 2 and VA 4 can be clearly distinguished from strain DTS 6 since aggregation does not occur in the former pair of strains but proceeds normally at the nonpermissive temperature in the latter strain. The final step in develop-

MUTANT STRAINS

ENZYMES		VA-5	DA 2	Agg 206	TS 2 or VA 4	DTS 6	KY 3	MIN 2
3	NAG	−	+	+	+	+	+	
4	MAN		+	+	+	+	+	
6	TD			−	+	+	+	
1	AT				−			+
2	LAP				−	+	+	+
5	TPS				−			+
7	TT			−	−	+	+	+
11	GP			−	−	−	+	
12	AIKP				−	−	−	+
13	βG-2				−	−	−	+

Fig. 9.12. Pattern of biochemical differentiations in morphological mutants. See legend to Fig. 9.1 for abbreviations of enzymes and number code. The morphological capabilities of the strains are given in Fig. 9.13. Accumulation of an enzyme was judged to be plus if at least 80% of wild-type peak activity was reached. Accumulation was judged negative if less than 20% of the normal increase occurred (Loomis, 1974).

MUTANT STRAINS

Stage	VA-5	DA-2	Agg 206	TS 2or VA 4	DTS 6	KY 3	MIN 2	Wh 1
Aggregate	−	−	−	−	+	+	+	+
Grex	−	−	−	−	−	+	+	+
Fruit	−	−	−	−	−	−	+	+
Spores	−	−	−	−	−	−	−	+
Pigment	−	−	−	−	−	−	−	−

Fig. 9.13. Pattern of morphological differentiations in morphological mutants. The various mutant strains have been described in the text. A plus indicates clear attainment of the indicated stage.

AGGREGATE → GREX → FRUIT → SPORES → PIGMENT

```
                        Agg      VA 4
            VA-5  DA 2   206    or TS 2  DTS 6   KY 3   MIN 2   Wh 1
            A ⇥ B ⇥ C ⇥ D ⇥ E ⇥ F ⇥ G ⇥ H ⇥ I
                         ① AT  ⑤ TPS      ⑪ GP   ⑫ AIKP        DTS 6
                         ② LAP ⑦ TT                            GI
                               ⑥ TD
                         ③ NAG
                         ④ MAN                                 ⑬ βG-2
```

Fig. 9.14. Genetic dissection of developmental steps in *Dictyostelium*. A through I are a dependent sequence of discrete steps defined by the mutant strains separating them. Arrows indicate the resultant differentiations. See legend to Fig. 9.1 for abbreviations of the enzymes. In a few places there is an uncertainty of one step in assigning the step leading to specific enzymes (Loomis, 1974).

ment is defined by the white strain, Wh 1, which fails to accumulate the carotenoid pigment.

When considered together these studies define 8 unequivocal steps in development which form a linear progression (Fig. 9.14). The intervening stages are referred to as stages A through I. At present we have no idea what processes or functions comprise these stages. The stage-specific enzymes themselves do not appear to be in the main causal progression since structural gene mutations in at least two of them, acetylglucosaminidase and α-mannosidase, do not modify subsequent biochemical differentiations (see Chapter 11).

A branch point appears to occur at stage G leading to the step necessary for accumulation of β-glucosidase-2. When strain DTS 6 is incubated at the nonpermissive temperature, pseudoplasmodia formation and further development is blocked and neither alkaline phosphatase nor β-glucosidase-2 accumulate (see Chapter 12). However, if cells of this strain are shifted down to the permissive temperature after 26 hours, morphogenesis proceeds again and normal fruiting bodies are formed. Under these conditions alkaline phosphatase accumulates normally but β-glucosidase-2 is not synthesized. These results indicate that a step specific to β-glucosidase-2 accumulation is irreversibly affected by nonpermissive conditions in strain DTS 6 while other steps are fully reversible in this strain.

Biochemical analysis of the morphological mutants has thus sketched out some of the underlying temporal controls in *Dictyostelium*. We have yet to gain direct evidence on the physiological role of most of the stage-specific enzymes. The morphological mutants shed light on their control but not on their function. This is discussed further in Chapter 11. It must be emphasized that there are likely to be several hundred enzymes which are orchestrated by the developmental program and only a few have been recognized so far. As more markers are analyzed in an expanded series of morphological strains the number of defined steps will increase and it is likely that further branch points will be recognized.

Chapter 10

Changes in Metabolites during Development

Cells Grown on Bacteria

Concomitant with changes in polysaccharides, RNA molecules, and proteins, the cellular concentrations of several small molecules change significantly during development of *D. discoideum*. Some of these changes can be accounted for as a consequence of the altered enzymatic activities while others appear to reflect the change from an exogenous to an endogenous energy source. Many of the metabolites studied are components of the central metabolic pathway of *D. discoideum* and are subject to interconversions catalyzed by a series of enzymes. The concentration of these compounds is a complex function of the metabolic flow rates through the various pathways. In each pathway a limiting reaction will determine the maximum flow along that pathway. A summary of the known interconversions of central metabolism in *D. discoideum* is presented in Fig. 10.1.

CENTRAL METABOLIC PATHWAYS OF DICTYOSTELIUM

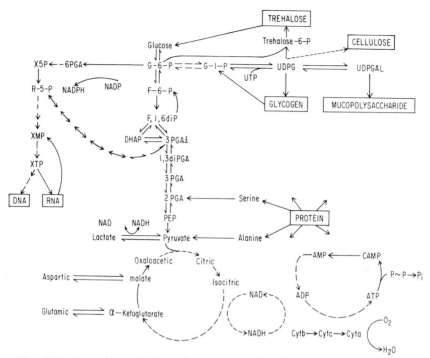

Fig. 10.1. Central metabolic pathways of *D. discoideum*. Most of the reactions and metabolites are common to all cells and the standard abbreviations used are for the intermediates of the Embden-Meyerhof and related pathways. Solid lines indicate some reactions which have been studied in *D. discoideum* and reported in the literature (see text). Major end products are emphasized in boxes.

Cyclic AMP plays a central role in chemotactic aggregation of *D. discoideum*. It has been reported that both the intracellular and the extracellular concentrations of cAMP increase more than 50-fold during the first 10 hours of development to 0.075 m*M* (Table 10.1) (Malkinson and Ashworth, 1973). The intracellular concentration subsequently drops during the later stages of development to less than 0.01 m*M*, but often rises again during culmination. The increase in cAMP that occurs shortly after the initiation of development may result from the inhibition of cAMP phosphodiesterase by the specific protein inhibitor (Riedel *et al.*, 1972). During the pseudoplasmodial stage, phosphodiesterase activity reappears as the inhibitor pool is depleted. This enzyme may be responsible for the drop in cAMP seen in fruiting bodies.

TABLE 10.1

METABOLITE CHANGES

Metabolite	Approximate intracellular concentration (μM)			Reference
	Amoebae	Pseudo-plasmodia	Fruiting bodies	
cAMP (intracellular)	1	75	10	Malkinson and Ashworth, 1973
cAMP (extracellular)	0.01	0.6	1.0	Malkinson and Ashworth, 1973
P_i	5000	10,000	50,000	Gezelius and Wright, 1965
ATP	—	900	1,400	Wright, 1963b
ADP	—	300	400	Wright, 1963b
AMP	—	100	80	Wright, 1963b
TPN	16	20	14	Wright and Wassarman, 1964
TPNH	26	22	27	Wright and Wassarman, 1964
DPN	33	27	25	Wright and Wassarman, 1964
DPNH	38	21	37	Wright and Wassarman, 1964
Glucose	60	300	60	Wright *et al.*, 1964
G-6-P	30	100	20	Wright *et al.*, 1964
UDPG	30	100	10	Wright *et al.*, 1964
G-1-P	3	10	2	Pannbacker, 1967b
Glutamate	100	200	1,000	Wright, 1963b
F-6-P	1.4	—	10	Cleland and Coe, 1969
F-1-6 diP	5	—	10	Cleland and Coe, 1969
DHAP	500	—	100	Cleland and Coe, 1969
Glyceraldehyde 3P	3	—	10	Cleland and Coe, 1969
2-P-glycerate	12	—	50	Cleland and Coe, 1969
Pyruvate	50	—	500	Cleland and Coe, 1969
Ubiquinone Q8,9	100	—	500	Long and Coe, 1973

The intracellular concentration of inorganic phosphate increases throughout development from an initial concentration of 5 mM to a peak of about 50 mM in fruiting bodies (Table 10.1) (Gezelius and Wright, 1965). Many reactions including several steps in protein synthesis liberate inorganic phosphate. Moreover, pyrophosphate is formed during RNA synthesis and in the synthesis of UDPG from G-1-P and UTP. Pyrophosphate is rapidly converted to inorganic phosphate by the action of the pyrophosphatase present throughout development.

Phosphate is usually recycled through oxidative phosphorylation to ATP and subsequently utilized in reactions such as the one catalyzed by glucokinase. However, the accumulation of inorganic phosphate during development indicates an imbalance between phosphate release and reutilization. It is possible that the cells store phosphate for utilization following germination. An exceptionally high concentration of phosphate has been found in spores (Gezelius and Wright, 1965).

There appears to be little change in the intracellular concentration of the adenine nucleotides or pyrimidine nucleotides during development (Table 10.1) (Wright, 1973b; Wright and Wasserman, 1964). These coenzymes play roles in a large variety of reactions and one would not expect dramatic changes in the intracellular concentrations.

Free glucose increases from a concentration of 0.06 to 0.3 mM during aggregation and then falls back to the initial level at culmination (Table 10.1) (Wright et al., 1964; White and Sussman, 1961). The increase early in development may reflect the action of amylases on stored glycogen as discussed in Chapter 8. Later in development glucokinase may mobilize glucose into polysaccharide synthesis.

The intracellular concentrations of glucose 6-phosphate, glucose 1-phosphate, and UDPG parallel those of glucose and may also reflect the initial rapid catabolism of stored polysaccharides followed by a period of increased biosynthetic activity during culmination (Table 10.1) (Wright et al., 1964; Pannbacker, 1967b). The increase in specific activity of UDPG pyrophosphorylase correlates roughly with the decrease in intracellular UDPG concentration.

The intracellular concentration of glucose 6-phosphate decreases during the period of trehalose synthesis. Formation of the storage disaccharide trehalose from glucose 6-phosphate may be one of the causes for depletion of this pool. Glucose 6-phosphate also acts as an effector of glycogen synthetase, increasing the affinity of the enzyme to UDPG. *In vitro* addition of glucose 6-phosphate at the concentration found in pseudoplasmodia, 0.1 mM, lowers the K_m for UDPG 2.5-fold to a value of 2×10^{-3} M (Wright et al., 1968). Since UDPG is present at less than saturating concentrations at all stages in development, the concentration of glucose 6-phosphate may be a factor in regulation of glycogen synthesis.

Amino acid catabolism proceeds throughout development and appears to be a major source of energy. The first step in catabolism of many amino acids is a transamination to α-ketoglutarate with the resulting formation of glutamate. The intracellular concentration of glutamate increases steadily throughout development reaching a final level of 1 mM (Table 10.1) (Wright, 1963b). It would appear that the increased

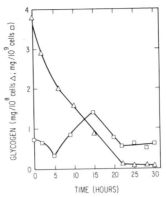

Fig. 10.2. Glycogen content during differentiations of amoebae of strain Ax2. Cells were grown so as to have either high (△———△) or low (□———□) initial glycogen content (Ashworth and Weiner, 1972).

flow of nitrogen metabolism through the glutamate pool exceeds the capacity of glutamic dehydrogenase to catalyze the deamination of glutamate. The enzyme remains at a constant specific activity throughout development.

There are no dramatic changes in the intracellular concentrations of most of the intermediates of the Embden-Myerhof pathway although there is a general trend toward somewhat larger pools of these compounds in developing cells (Table 10.1) (Cleland, 1969). The most marked increase in the intracellular concentration of these compounds is that of pyruvate which increases 10-fold to a final level of 0.5 mM. This may reflect an increase in metabolism of amino acids leading to pyruvate.

The intracellular concentration of total neutral lipids of *D. discoideum* changes little during development. However, a specific neutral lipid fraction has been found to increase during culmination. This fraction contains the ubiquinones, Q8 and Q9, with the Q8 molecule predominating (Table 10.1) (Long and Coe, 1973). It is not known what role these compounds may play in metabolism or development of *D. discoideum*.

Axenically Grown Cells

Cells grown axenically in medium containing glucose have a greatly increased glycogen content as is discussed in Chapter 8. This has been found to profoundly alter the pattern of intracellular metabolites during

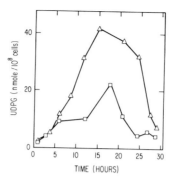

Fig. 10.3. UDP glucose content during differentiation of amoebae of strain Ax2 containing initially 7.8 mg glycogen/10^8 cells (△——△) or 0.26 mg glycogen/10^8 cells (□——□). [Adapted from Garrod and Ashworth (1973). Development of the cellular slime mould *Dictyostelium discoideum. Symp. Soc. Gen. Microbiol.* **23,** 407–435. With permission of Cambridge University Press.]

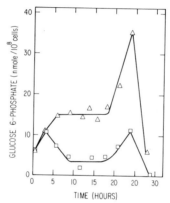

Fig. 10.4. Glucose 6-phosphate content during differentiation of amoebae of strain Ax2 containing initially 7.8 mg glycogen/10^8 cells (△——△) or 0.26 mg glycogen/10^8 cells (□——□). [Adapted from Garrod and Ashworth (1973). Development of the cellular slime mould *Dictyostelium discoideum. Symp. Soc. Gen. Microbiol.* **23,** 407–435. With permission of Cambridge University Press.]

development (Garrod and Ashworth, 1973). Cells grown axenically in the absence of glucose have glycogen levels comparable to those of bacterially grown cells. Most of the glycogen stored in cells grown on glucose is removed during the aggregation and pseudoplasmodial stages, and the final glycogen content is similar to that of bacterial grown cells (Fig. 10.2). Cells with an initially high glycogen content accumulate UDPG to a level almost twice as high as that in cells with an initially low glycogen content (Fig. 10.3). The difference may reflect a decreased

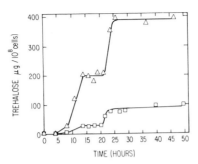

Fig. 10.5. Trehalose content during differentiation of amoebae of strain Ax2 containing initially 6.0 mg glycogen/10⁸ cells (△———△) or 0.3 mg glycogen/10⁸ cells (□———□). [Adapted from Garrod and Ashworth (1973). Development of the cellular slime mould *Dictyostelium discoideum. Symp. Soc. Gen. Microbiol.* **23,** 407–435. With permission of Cambridge University Press.]

demand for glycogen synthesis in those cells which already have a high glycogen content or may reflect an increase in glucose 1-phosphate production by the action of glycogen phosphorylase. An increase in the glucose 1-phosphate pool would stimulate UDPG formation.

The concentration of glucose 6-phosphate is also higher during development in cells with an initially high glycogen content (Fig. 10.4). It rises to a peak during culmination almost 3 times higher than that found in cells which started development with a low glycogen content.

The consequences of these increases in metabolites can be seen in the pattern of trehalose accumulation (Fig. 10.5). The disaccharide accumulates to a level about 3-fold higher in cells with a high initial glycogen content. Trehalose-phosphate synthetase activity follows the same developmental kinetics in cells with either a high or low initial glycogen activity and accumulates to the same specific activity (Garrod and Ashworth, 1973). Thus, the increase in trehalose synthesis in cells with a high content of glycogen does not appear to be a consequence of increased enzymatic activity but rather the result of increased intracellular concentrations of the precursors UDPG and glucose 6-phosphate.

At present we have information on the intracellular concentration of only a few of the key intermediates in the metabolism of *D. discoideum.* Undoubtedly there are significant changes in many more metabolites. Together with changes in the enzymatic complement of the cells, the concentration of small molecules will determine the relative rate of flow down each pathway at various stages. Once we have a large body of information on the changes in metabolites under a variety of developmental conditions it may become clear how each component interacts to produce the overall physiological pattern.

Chapter 11

The Relation of Biochemical
Differentiations to Morphogenesis

Profound metabolic changes underlie the morphological stages observed during development of *D. discoideum*. Yet we know very little about how these changes direct morphogenesis or even if they are essential for the observed cytodifferentiations. However, the relation among specific gene expression, altered cellular physiology, and morphogenesis is one of the most fascinating aspects of developmental biology. Although we know only a few pieces of this puzzle, it is interesting to try to position them in an attempt to account for the known changes in shape.

Aggregation

Aggregation results in the piling up of a large number of cells in groups essential for subsequent morphogenesis. As discussed in Chapter 5, this first step in the development of *D. discoideum* is mediated by

139

chemotaxis to cAMP. The attractant is released into the environment during the stage when a specific inhibitor of cAMP phosphodiesterase blocks the extracellular enzyme. However, chemotaxis is not sufficient by itself to form an aggregate. Cellular cohesion must increase to produce the tight mass of cells. Cohesion may be mediated in this system by a carbohydrate-binding protein which increases dramatically at this stage (Rosen *et al.*, 1973). Thus, changes in a few known macromolecules may account for the radical change in cell behavior immediately following the initiation of development.

Pseudoplasmodium Formation

Subsequent morphogenesis is not as easy to account for. The surface sheath appears to confer integrity to the mass of cells and control phototactic migration. However, we know very little of the detailed structure or biosynthesis of the sheath components. Nor do we have direct measurements of the relative affinity of cells for sheath or each other.

One of the few structures which can be seen to accumulate during the pseudoplasmodial stage in preparation for culmination are the prespore vesicles. These organelles appear to form the outer case of the spores when terminal differentiation takes place. They contain material similar to the mucopolysaccharide found in sori. The enzymes which are likely to catalyze the synthesis of this mucopolysaccharide appear at a distinct stage; both UDPgalactose epimerase and UDPgalactose polysaccharide transferase accumulate at about 18 hours of development at the same time as the mucopolysaccharide appears in the vesicles. The appearance of these enzymes may control accumulation of this polysaccharide. Eversion of the prespore vesicles followed by polymerization of the contents could play an essential role in morphogenesis of spores.

Culmination

Much of the final shape of the fruiting bodies of *D. discoideum* is determined by the stalk which initially skewers the posterior mass of pseudoplasmodial cells and then lifts the mass into the air. The stalk sheath contains a high content of cellulose and thus the control of cellulose synthesis may be a critical event in determining the time and place of stalk formation. Cellulose is seen to accumulate dramatically late in development (Fig. 8.5). However, it is important to remember that the polymerization of cellulose in the extracellular space of the stalk may involve several variables other than cellulose synthesis such as secretion and fiber formation. Hopefully, we will soon learn more about such processes.

Meanwhile, we can consider the steps which may affect the synthesis of cellulose from UDPG. The reaction is catalyzed by cellulose synthetase, but we have few hints on how this enzyme might control stalk formation since few studies have been done on the kinetics of accumulation of this enzyme during development. The *in vivo* concentration of UDPG appears to change during development reaching a peak of 8×10^{-4} M just before culmination (Table 10.1; Fig. 10.3). It is conceivable that the period of cellulose deposition is controlled in part by the internal concentration of UDPG. However, the concentration of UDPG appears to decrease before the period of cellulose synthesis and thus cannot directly account for the increased rate of synthesis. Clearly there must be further controlling effects which stimulate cellulose synthesis during the period of stalk formation.

Glycogen Metabolism

Measurement of glycogen synthesis in relation to morphogenesis is complicated by the presence of two pools of glycogen in *D. discoideum* (see Chapter 8). Moreover, the total glycogen content can vary enormously without affecting morphogenesis. Nevertheless, considerable speculation has centered on glycogen metabolism as a criterion of differentiation in *Dictyostelium* (Wright, 1973; Hames *et al.*, 1972).

The rate of glycogen synthesis is low during aggregation and increases about 3-fold during the pseudoplasmodial stage to a peak of 120 μg glucose incorporated per hour per 10^8 cells (Fig. 8.6). Thereafter the rate decreases sharply. The specific activity of glycogen synthetase appears to stay essentially constant throughout development so that changes in catalytic potential cannot account for the pattern of glycogen synthesis. However, the concentration of the substrate, UDPG, may affect the rate of glycogen synthesis. The concentration of UDPG, which is limiting at all stages, increases from 3- to 10-fold between the aggregation and pseudoplasmodial stage (Table 10.1; Fig. 10.3). The concentration drops back to the original low value during culmination. Moreover, the affinity of the enzyme for UDPG is modified by glucose 6-phosphate which also accumulates prior to culmination (Table 10.1; Fig. 10.4). The combined effect of increased levels in UDPG and glucose 6-phosphate may account for the increase in rate of glycogen synthesis.

Substrate Concentrations

UDPG serves as a precursor of a variety of carbohydrates which accumulate late in development. Thus, the availability of UDPG for glycogen synthesis is not independent of the rate of cellulose, mucopolysaccharide, and trehalose synthesis. Mechanisms appear to exist to balance the synthesis of UDPG with its utilization in the various pathways since the internal pool of UDPG is never depleted. The flow of molecules down the various pathways must be balanced in a complex manner by the relative catalytic functions of the enzymes and their affinities for the substrates.

During the pseudoplasmodial stage the specific activity of UDPG pyrophosphorylase increases about 10-fold due to an increase in the differential rate of synthesis (Franke and Sussman, 1973). The cellular concentrations of the precursors of UDPG, glucose 1-phosphate, and UTP, also increase during the pseudoplasmodial stage (Table 10.1). Since these compounds are present at less than the K_m of the enzyme, the increased concentrations together with the increased specific activity of the enzyme should result in a 20- to 30-fold increase in UDPG synthesis. This may be sufficient to accommodate the demand for UDPG during culmination.

Glucose 1-phosphate, in turn, is produced both by gluconeogenesis and by glycogen degradation. Shortly after the initiation of development, storage glycogen appears to be broken down by the action of amylase. Later in development the accumulation of glycogen phosphorylase may ensure that sufficient glucose 1-phosphate is available for UDPG formation.

UDPgalactose epimerase appears just before culmination and directs the conversion of UDPG to UDPgalactose which is then incorporated into mucopolysaccharide. Accumulation of both of these enzymes at culmination will draw some of the UDPG into this pathway. Trehalose also accumulates during culmination and is synthesized from UDPG and glucose 6-phosphate in a reaction catalyzed by trehalose-phosphate synthetase. This pathway also places demands on available UDPG.

Metabolic Modeling

Wright and her colleagues (Wright, 1973; Wright and Gustafson, 1972) have attempted to analyze the interaction of the pathways leading to and from UDPG (Fig. 11.1). Data obtained on the cellular concentra-

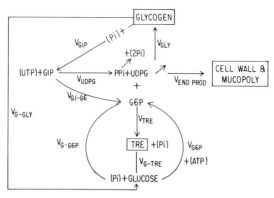

Fig. 11.1. Carbohydrate metabolism used in kinetic model simulation of end product synthesis during differentiation of *Dictyostelium*. TRE refers to trehalose while other abbreviations have been used previously. V refers to the actual *in vivo* rate of indicated reaction. The intermediate arrow from UDPG to glycogen or cell wall and mucopolysaccharide indicates a proposed channeling of carbohydrate metabolism during culmination (Wright and Gustafson, 1972).

tions of the intermediates were supplied to a computer programmed with the estimated *in vivo* reaction rates and the *in vitro* kinetic parameters of the enzymes of this model. Various parameters were changed so that the enzyme activities and metabolite levels observed *in vivo* were mimicked by the computer. A successful solution required: (1) a closed cycle prior to culmination in which glycogen degradation (V_{G-1-P}) and glycogen synthesis (V_{GLY}) are balanced; (2) a 3-fold increase in the rate of glycogen degradation (V_{G-1-P}) in a linear fashion from aggregation to culmination; (3) an abrupt cessation of glycogen synthesis (V_{GLY}) at culmination and an equally abrupt increase in cell-wall polysaccharide, mucopolysaccharide, and trehalose synthesis ($V_{END\ PROD}$; V_{TRE}) at this time.

A metabolic model constructed with these components has successfully accounted for the changes in several carbohydrates which occur during development (Wright and Gustafson, 1972). However, this analysis is by necessity a simplification of the true situation. It does not take into account the active gluconeogenesis during development that is discussed in Chapter 8, and evidenced by the fact that carbon of labeled glutamate enters the pool of cellulose at least 25% as well as carbon from ^{14}C-glucose (Wright and Bloom, 1961). Thus, the present model attempts to account for the observed changes in cellulose, mucopolysaccharide, and trehalose without consideration of net flow into the carbohydrate pool. The analysis also assumes that total glycogen remains constant in the cells until culmination when it drops precipitously. As discussed in Chapter 10, the initial glycogen content varies considerably depending on prior growth conditions. In the specific case that was considered, that of bacterially grown cells, glycogen was found to increase severalfold during the pseudoplasmodial stage and then to fall during culmination (Fig 8.5). Thus, a critical assumption on which the present model is built does not appear to reflect the experimental observations. The model does not take into consideration the more recent evidence that 2 pools of glycogen may exist. Perhaps for these reasons the model predicted that glycogen synthesis should decrease only at culmination, while the observed kinetics of glycogen synthesis (Fig. 8.6) show a decrease about 6 hours earlier. The abrupt increase in end product synthesis most likely results from accumulation of the pertinent enzymes since gradual changes in substrate concentrations cannot account for the rapid accumulation of end products during discrete periods. However, Wright (1973) has suggested that enzyme accumulation may not be a critical step and has postulated a gate functioning to shunt the flow of glucose 1-phosphate from glycogen synthesis to cell wall, mucopolysaccharide, and trehalose synthesis. The biochemical basis for such a switching mechanism is not defined.

Although we can account for the flow of UDPG into mucopolysaccharide by the sudden appearance of UDPgalactose epimerase and UDPgalactose transferase, we still cannot account from an enzymatic point of view for the sudden flow of UDPG into cellulose. Nor does the abrupt appearance of trehalose conform to the known developmental kinetics of trehalose-phosphate synthetase. We are left with the possibility that unknown factors are functioning in this system which are essential to the redirection of metabolic flow down the various pathways.

Computer analysis of metabolism is potentially a powerful approach to developing systems in which each individual metabolic step is well characterized but the overall complexity of the system precludes a simple analysis. It is important to recognize some of the requirements of this approach:

1. The *in vitro* characteristics of the various enzymes must give an accurate picture of the *in vivo* enzyme characteristics.

2. Effectors of a reaction, both small molecules and macromolecules, must be recognized and accurately measured.

3. The intracellular concentration of metabolites must be accurately determined and cellular compartmentalization must be taken into account.

4. The flow pattern delineated by the model must include all pertinent enzymatic steps within the area of metabolism as well as contributing pathways.

Many of these requirements are exceptionally difficult to fulfill. The preliminary modeling which is being done now clearly does not satisfy these conditions and may give an erroneous impression that a unique solution can be reached with the data available. As further data becomes available, more complex models may lead to insights on the flow of metabolites to end products specific to differentiated cells.

Although detailed modeling may be premature in *D. discoideum*, it has focused attention on the critical role that the concentration of metabolites may play in directing physiology in a differentiating system (Wright, 1973). The pathways to the major polysaccharides lead out from common precursors, and it is essential to consider changes in catalytic activities in context with other changing parameters. The concentrations of almost all metabolites are considerably below the K_m's of the pertinent enzymes and a fall in the concentration of a specific compound will decrease the flow down a variety of pathways. For instance, if the rate of production of glucose 1-phosphate decreased below the rate of its conversion to UDPG, then the pool of glucose 1-phosphate would decrease until the rate of UDPG synthesis was brought into balance as a result of substrate limitation. In other words, a linear metabolic pathway cannot function more rapidly than the slowest step. A decrease in the UDPG pool would affect the pathways to various end products differently depending on the affinity of the enzymes for UDPG. However, competition between the pathways will also be determined by the relative activities of the enzymes.

Until most of the significant variables are known in even as simple a system as polysaccharide formation in *D. discoideum*, it will be exceptionally difficult to establish the relative role of each component in metabolic flow patterns and morphogenesis. An alternative approach is to perturb the system by well-defined mutations in the genes coding for the pertinent enzymes. Mutants carrying enzymes with altered substrate affinity can indicate dependence on substrate concentrations while temperature-sensitive and absolute mutations can delineate the significance of the overall metabolic pathway.

Genetic Analysis

A series of mutations affecting N-acetylglucosaminidase has recently been isolated (Dimond *et al.*, 1973). Six strains were found which formed less than 1% of this enzyme activity. Four other strains formed thermolabile enzyme while one strain formed N-acetylglucosaminidase which had an altered substrate affinity. The temperature-sensitive strains clearly carry mutations in the structural gene for N-acetylglucosaminidase and can be studied under permissive and nonpermissive conditions.

The absolute mutants develop on filter supports in an apparently normal manner and form well-shaped fruiting bodies. The stage-specific enzymes accumulate at the normal times and to the normal extent in these strains. However, under migration conditions one can observe the effects of loss of N-acetylglucosaminidase. The cells aggregate on agar but cannot maintain normal pseudoplasmodial size during migration; the rate of migration is also greatly reduced. This phenotype is observed only at the nonpermissive temperature in the temperature-sensitive strains and undoubtedly is a direct consequence of the loss of the enzyme.

Mutations affecting other enzymes of *D. discoideum* development will help delineate their function in directing morphogenesis. Isolation techniques are now being developed which appear to be applicable to several of the stage-specific enzymes and it is only a matter of time before we should have direct evidence of the interactions of these enzymes and their relation to morphogenesis.

Chapter 12

Modifications of Development

A great variety of physical and chemical conditions have been found which affect morphogenesis of *D. discoideum*. In all but a few cases the primary process which is modified is unknown and only the ultimate consequences can be assessed. Yet as our understanding of the critical physiology underlying development increases, these phenomena may be helpful in recognizing biochemical interactions. For this reason I have attempted in this chapter to catalog some of the ways in which development can be perturbed in *D. discoideum*.

Submerged Cultures

Dictyostelium develops normally only at an air–water interface (Gerisch, 1968). Submerged cells aggregate and form tight cell groups but do not undergo further development underwater. Cellular cohesion increases in submerged cultures and leads to well-formed aggregates (Beug and Gerisch, 1972). However, pseudoplasmodial polarity is never established in submerged conditions and surface sheath is not formed around the aggregates. Neither spores nor stalks are formed in submerged cultures.

Cell Density

When aggregation proceeds at an air–water interface, such as on the surface of an agar plate, there is a certain minimal density of cells below which the cells fail to aggregate (Bradley et al., 1956). If vegetative cells of D. discoideum are spread at a density of less than 80 cells per mm^2, no aggregates are formed. A density of 200 cells per mm^2 results in maximal aggregative performance. Densities greater than 1000 cells per mm^2 often result in aggregates which are initially larger than those formed at lower densities but these break up during or after pseudoplasmodium formation to result in normal sized pseudoplasmodia containing about 10^5 cells. The final size of pseudoplasmodia may depend on the physical properties of the sheath and the conditions of migration.

Aggregation is unaffected by a wide range of ionic conditions and proceeds well between pH 5.5 and 8.5 (Bradley et al., 1956). On the other hand, histidine at 10^{-2} M has been found to reduce the minimal population density required for aggregation more than 2-fold (Bradley et al., 1956). The biochemical basis for this effect is unknown but may be related either to cAMP production or sensitivity of the cells to the attractant.

Overhead Light

High ionic strength and the presence of overhead light induce culmination of pseudoplasmodia (Newell et al., 1969). Aggregates which form in an environment of low ionic strength migrate for extended periods. When they are given overhead light or shifted to an environment of high ionic strength, they stop migrating and begin forming a stalk. Overhead light elicits vertical phototaxis which may limit further migration. A drop in humidity also induces culmination (Bonner and Shaw, 1957). The normal developmental program of accumulation of the stage-specific enzymes presented in Chapter 9 was determined in cells developing in an environment of fairly high ionic strength. Migration of pseudoplasmodia does not occur under these conditions. The biochemical program is altered when development proceeds in a low ionic environment, and the cells are induced to migrate. UDPG pyrophosphorylase and UDPgal-

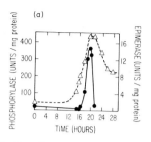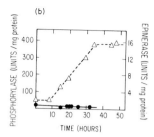

Fig. 12.1. Developmental kinetics of UDPG pyrophosphorylase ($\triangle$——$\triangle$) and UDPgalactose epimerase ($\bullet$——$\bullet$) in cells developing (a) on filters soaked in buffered salt solution; (b) on unbuffered agar (Ellingson *et al.*, 1971).

actose transferase accumulate under these conditions but UDPgalactose epimerase is not formed (Fig. 12.1) (Ellingson *et al.*, 1971). When the cells are shifted to a high ionic environment with overhead light, epimerase accumulates within a few hours and reaches essentially the same specific activity seen under standard conditions (Fig. 12.2). Accumulation of epimerase after light induction has been shown to require concomitant RNA synthesis (Ellingson *et al.*, 1971).

Accumulation of UDPG pyrophosphorylase occurs in the dark in a low ionic environment but takes 40 hours to reach maximal activity rather than the usual 20 hours (Fig. 12.1) (Newell and Sussman, 1970). If culmination is induced with overhead light during the period of slow accumulation the rate increases rapidly and peak activity is reached within a few hours (Fig. 12.3). If, on the other hand, culmination is induced after peak activity has been reached, a period of further synthesis is induced and the specific activity increases to about 650 units

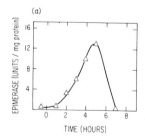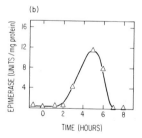

Fig. 12.2. Developmental kinetics of UDPgalactose epimerase following a shift to well-buffered salt solution and overhead illumination. Pseudoplasmodia were allowed to migrate from unbuffered agar to filter supports. At the time indicated the filters were shifted to buffered salt solution and given overhead light. The pseudoplasmodia had migrated either 10 mm (a) or 50 mm (b) before being shifted (Ellingson *et al.*, 1971).

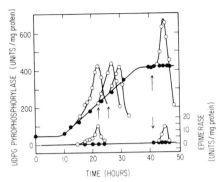

Fig. 12.3. Developmental kinetics of UDPG pyrophosphorylase following a shift to overhead light. Pseudoplasmodia migrating on agar were induced to culminate by overhead light at the indicated times (O——O). Control cells were kept in the dark (●——●) (Newell and Sussman, 1970).

(Newell and Sussman, 1970). It has been suggested that the second period of accumulation results from the synthesis of a "quantum" of enzyme (Sussman and Newell, 1972). A quantum is defined in this context as the amount of enzyme which accumulates under the normal high ionic conditions (i.e., 450 units/mg protein for UDPG pyrophosphorylase).

Dissociation

Cells can be mechanically dissociated from aggregates or pseudoplasmodia at any stage of development with little harm to the cells. When redeposited, the cells reaggregate rapidly and return to the stage from which they were dissociated (Loomis and Sussman, 1966). Dissociation of pseudoplasmodia at 14 hours of development does not affect accumulation of any of the stage-specific enzymes. However, dissociation at later times results in an increase in the peak specific activity of several stage-specific enzymes (Newell *et al.,* 1972). By 16 hours trehalose-6-phosphate synthetase has almost reached peak activity (10 units/mg protein). Dissociation at this time results in renewed synthesis and the peak activity increases to 17 units/mg protein (Fig. 12.4). The accumulation of both UDPG pyrophosphorylase and UDPgalactose transferase responds to dissociation in a similar manner (Figs. 12.5 and 12.6). Epimerase activity initially drops following dissociation but then increases. Repeated disaggregation results in a further increase in specific activity of these enzymes (Fig. 12.7). The increase in specific activity requires RNA synthesis following disaggregation (Newell *et al.,* 1972).

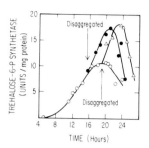

Fig. 12.4. Accumulation of trehalose-6-phosphate synthetase activity after disaggregation at 15.5 hours (●——●) or 19 hours (△——△). Undisturbed control cells (○——○) (Newell, Franke and Sussman, 1972).

Disaggregation and subsequent reaggregation during the pseudoplasmodial stage does not significantly modify the developmental kinetics of N-acetylglucosaminidase, α-mannosidase, or β-glucosidase-2 (Newell

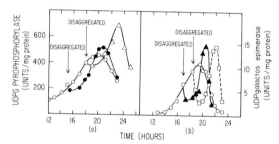

Fig. 12.5. Accumulation of UDPG pyrophosphorylase and UDPgalactose epimerase after disaggregation. Cells were disaggregated at the indicated times and assayed either for UDPG pyrophosphorylase (left) or UDPgalactose epimerase (right). Undisturbed controls (○——○); disaggregation at 15 hours (●——●), 18 hours (△——△); 17 hours (▲——▲); 18.5 hours (□——□) (Newell *et al.,* 1972).

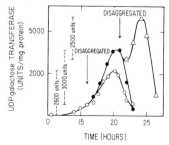

Fig. 12.6. Accumulation of UDPgalactose transferase following successive disaggregations at the times indicated. Undisturbed controls (○——○); disaggregated at 16 hours (●——●); and again at 22 hours (△——△).

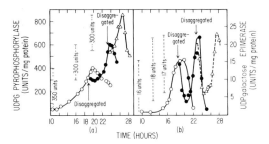

Fig. 12.7. Successive disaggregations were performed at the times noted and the cells were redeposited on fresh filter supports. UDPG pyrophosphorylase (a); UDPgalactose epimerase (b). Undisturbed controls (O———O); disaggregated at 19 hours (●———●); and again at 23 hours (△———△) (Newell *et al.*, 1972).

et al., 1972, Loomis, 1970a, Coston and Loomis, 1969). Cells dissociated and allowed to reaggregate at 18 hours fail to accumulate alkaline phosphatase to the normal peak activity (Fig. 12.8) (Loomis, 1969c). It appears that dissociation irreversibly blocks accumulation of this late enzyme.

Sequential dissociation and reaggregation result in increased activity of UDPG pyrophosphorylase, epimerase, and transferase. Each round of synthesis following dissociation results in the accumulation of approximately the same amount of activity as seen in the original increase (Figs. 12.5 and 12.7). Sussman and Newell (1972) have suggested that dissociation triggers processes which result in accumulation of a quantum of activity. They consider an inductive event resulting in a programmed

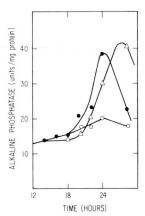

Fig. 12.8. Accumulation of alkaline phosphatase after disaggregation at 14 hours (△———△) or 18 hours (O———O). Unmolested controls (●———●) (Loomis, 1969c).

amount of transcription and a regulated amount of translation from each new mRNA molecule. The biochemical mechanisms which could elicit such behavior are unknown.

Another enzyme, glycogen phosphorylase, also accumulates following dissociation but reaches an only slightly higher specific activity (Fig. 12.9) (Firtel and Bonner, 1972b). In this case a quantum of activity is 24 units while only about 11 units accumulated after dissociation. The quantum rule does not seem to hold for this enzyme.

Dissociation and reaggregation clearly has a dramatic effect on the biochemical differentiation of *D. discoideum*. The accumulation of some enzymes responds by a new round often equalling the normal first round; in others, less than the normal amount accumulates. Dissociation and reaggregation obviously interrupts the multicellularity of the cell masses and removes diffusable molecules generated within the mass. These modifications appear to be variously interpreted by the individual genes of the stage-specific enzymes and result in the varied pattern of response.

By replating cells dissociated from pseudoplasmodia at a low concentration, reaggregation can be prevented. If cell contact is not reestablished following dissociation neither UDPG pyrophosphorylase nor UDPgalactose transferase accumulate (Fig. 12.10) (Newell *et al.*, 1971). The specific activity of UDPG pyrophosphorylase and UDPgalactose transferase remain constant under these conditions. Thus cells of *D.*

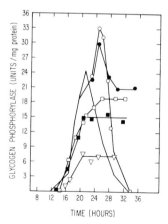

Fig. 12.9. Accumulation of glycogen phosphorylase after disaggregation. Pseudoplasmodia were disrupted and replated in the presence of 1.5×10^{-2} M EDTA. Dissociation was performed at 4 hours ($\triangledown$----$\triangledown$); 8.5 hours ($\blacksquare$——$\blacksquare$); 12 hours ($\square$——$\square$); 19 hours ($\bullet$——$\bullet$). Cells were also dissociated at 19 hours and allowed to reassociate in the absence of EDTA ($\bigcirc$——$\bigcirc$) (Firtel and Bonner, 1972b).

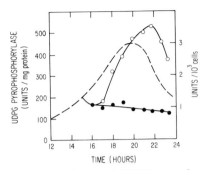

Fig. 12.10. Inhibition of accumulation of UDPG pyrophosphorylase in disaggregated cells not allowed to reaggregate. After 15 hours of development cells were dissociated and replated at either 4×10^6 cells/cm^2 (O———O) or at 2×10^3 cells/cm^2 (●———●). At the lower population density reaggregation did not occur. The values of UDPG pyrophosphorylase were determined relative to total protein or total cells that were equated at 16 hours (Newell *et al.*, 1971).

discoideum appear to require a multicellular environment for synthesis of these stage-specific enzymes. Attempts to biochemically define the necessary conditions have not been successful so far but are continuing in hopes of recognizing the signals involved in multicellular development of this organism.

Mechanical Barriers

Insertion of a small impermeable barrier part way into an early non-migrating pseudoplasmodium can lead to the formation of 2 tips. When these develop the mass splits and forms 2 independent fruiting bodies (Farnsworth, 1973b). If the barrier is inserted only slightly into a pseudoplasmodium the cells avoid it or push it away.

Partially inserted barriers can be removed at various times and the time for specification of polarity determined. Removal of the barrier less than 30 minutes after its insertion results in little modification of morphogenesis, while removal of the barrier after 40 minutes always results in 2 independent fruiting bodies. It appears that a stable developmental axis is formed in pseudoplasmodia 40 minutes after the system is perturbed (Farnsworth, 1973b).

Culmination and migration are blocked by partial insertion of a barrier and in some cases new tips cannot be observed. The mass of cells then differentiate directly into spores with little or no stalk formation. Terminal differentiation of spore cells appears to have been triggered in this case when movement relative to the sheath was blocked.

Terminal differentiation into stalk cells can be induced independently of spore differentiation by inserting nitrocellulose tubes into pseudoplasmodia (Farnsworth, 1973b). The enclosed cells lay down a heavy cellulose wall and vacuolize in a fashion characteristic of stalk cells. Nitrocellulose tubes will elicit stalk differentiation when inserted into any part of a pseudoplasmodium including the posterior portion in which cells would normally all make spores.

Effect of Nutrients

Removal of exogenous nutrients induces the cells to start the developmental stage of their life cycle. Growth medium can be added back to cells developing on filter supports by shifting the filters to pads saturated with HL-5 medium. If this is done any time during the first 4 hours of development, aggregation of axenic strains but not of strain NC-4 is delayed at least 24 hours (Loomis, unpublished). It appears that an exogenous food source keeps the cells in the growth phase and that an air–water interface is not a sufficient stimulus to induce development. When the components of HL-5 were tested individually at the concentration present in the medium it was found that neither glucose nor yeast extract delayed development. Proteose peptone, on the other hand, delayed aggregation when present by itself in the supporting medium. The effect could be observed only if proteose peptone had been present during the first few hours following deposition of the cells onto filter supports; thereafter, proteose peptone had no effect on morphogenesis.

Various other compounds can act singly to modify morphogenesis at various stages and many of these produce phenocopies of the mutant strains described in Chapter 5.

Phenocopies of Aggregateless Mutants

Aggregation can be inhibited by a variety of chemical treatments. On a crude level, replacement of the normal atmosphere by pure N_2 blocks all aggregation since development is an obligatory aerobic process (Bradley *et al.*, 1956).

High concentrations of cAMP in the environment likewise inhibit aggregation probably by swamping out the gradient of the attractant (Bonner *et al.*, 1969). The response is similar to that seen in mutant strains which lack phosphodiesterase and so cannot rid the local environment of accumulated cAMP.

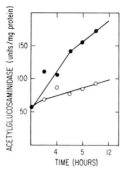

Fig. 12.11. Effect of L-canavanine on the accumulation of N-acetylucosaminidase. Amoebae were deposited on filters in the presence (O——O) or absence (●——●) of 1 mg/ml L-canavanine (Loomis, unpublished).

Ethylenediaminotetraacetetic acid (EDTA) at a concentration of 10^{-3} M in the environment also precludes aggregation, and keeps the cells as a smooth lawn of amoebae (Gerisch, 1961b). EDTA is a powerful chelator of divalent cations and it appears that Ca^{2+} ions are necessary for aggregation and cohesion.

Canavanine is an analog of arginine which is incorporated into proteins and disrupts their normal configuration. When 10^{-2} M canavanine is added at the time of initiation of development, aggregation is completely blocked. Protein synthesis is reduced to about 50% and the early stage-specific enzyme, N-acetylglucosaminidase, accumulates at a greatly reduced rate (Fig. 12.11) (Loomis, unpublished). When canavanine is added to pseudoplasmodia which have developed for 14 hours, it blocks terminal differentiation of spores, producing phenocopies of some fruitless mutants. The accumulation of UDPgalactose polysaccharide transferase is inhibited in cells treated with this analog (Fig. 12.12). The compound has also been found to block germination when added to washed spores (Cotter and Raper, 1970).

Phenocopies of Fruitless Mutants

Ethionine is an analog of methionine which results in abortive aggregates when present in the environment at 10^{-2} M. The cells form only rounded mounds reminiscent of the fruitless mutants (Filosa, 1960). The effect can be reversed by equimolar concentrations of methionine. Ethionine undoubtedly affects many biosynthetic functions which may effect transformation into pseudoplasmodia.

It has been reported that 10^{-3} M cAMP will stimulate a few cells

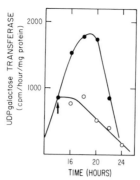

Fig. 12.12. Effect of L-canavanine when added 14 hours after the initiation of development. The specific activity of UDPgalactose polysaccharide transferase was determined in control cells (●———●) and in cells treated with the analog at 14 hours (○———○) (Loomis, unpublished).

placed at low density to form the heavy cellulose walls and large vacuoles characteristic of stalk cells (Bonner, 1970). The percentage of such cells is low and so no biochemical analysis of the phenomenon has been possible. Exogenous cAMP has yet other effects when presented to cells in the pseudoplasmodial stage (Nestle and Sussman, 1972). Addition of 3 mM cAMP at 16 hours of development leads to aberrant structures in which the majority of the cells remain at the base of stalks. Only about 10% of the cells at the base transform into spore cells. Addition of cAMP at later times has no effect on morphogenesis. When cAMP is added to migrating pseudoplasmodia just prior to culmination, the pseudoplasmodia form multiple fruiting bodies (Nestle and Sussman, 1972).

Addition of cAMP, at 16 hours, to cells developing on filter supports has no effect on the accumulation of UDPG pyrophosphorylase but inhibits further synthesis of epimerase (Fig. 12.13) (Nestle and Sussman, 1972). cAMP inhibits the second period of UDPG pyrophosphorylase accumulation elicited by the induction of culmination in migrating pseudoplasmodia by overhead light. It also blocks accumulation of epimerase under these conditions (Fig. 12.14). Clearly, cAMP has a profound inhibitory effect on late biochemical differentiations.

Phenocopies of Culmination Mutants

Several different chemical treatments result in phenocopies of the syndrome observed in mutant strain KY-3. The cells form normal pseu-

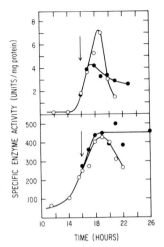

Fig. 12.13. Effect of cAMP on epimerase and UDPG pyrophosphorylase. After 16 hours development cells were moved to fresh salt solution (○——○) or to salt solution containing 3m*M* cAMP (●——●). Upper: epimerase; lower: UDPG pyrophosphorylase (Nestle and Sussman, 1972).

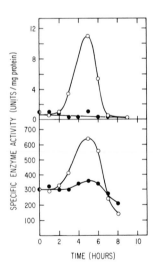

Fig. 12.14. Inhibition of accumulation of epimerase and UDPG pyrophosphorylase by cAMP. Pseudoplasmodia migrating on agar were transfered on filters to buffered salt solution and given overhead illumination (○——○). Half of the filters were transfered to buffered salt solution containing 0.3 m*M* cAMP and given overhead illumination (●——●). Upper: epimerase; lower: UDPG pyrophosphorylase (Nestle and Sussman, 1972).

doplasmodia but fail to culminate. An environment of low ionic strength will produce this response (Slifkin and Bonner, 1952; Newell *et al.*, 1969). Likewise treatment with colchicine will block culmination (Firtel and McMahon, personal communication). It is not known if this drug is acting in *D. discoideum* to block microtubule assembly as it does in other systems, but the phenomenon suggests it may be worthwhile to consider the role of microtubules in culmination.

The presence of 10^{-2} *M* adenine inhibits culmination and blocks both spore and stalk differentiation. The effect of adenine can be reversed by equimolar inosine but neither guanosine nor cAMP will reverse the effect. Accumulation of a late enzyme, alkaline phosphatase-2, is somewhat delayed by the presence of 10^{-2} *M* adenine but reaches peak specific activity by 36 hours.

L-Arginine at 3×10^{-2} *M* concentration leads to abnormal fruiting bodies in which the proportion of spores to stalk cells is greatly increased. The terminal structures appear as rounded white masses barely held off the support. The spore cells are regular and refractile but more rounded than normal.

Although L-cysteine does not affect the morphological aspects of culmination or encapsulation of spores, it almost completely inhibits the accumulation of the late stage-specific enzyme, alkaline phosphatase-2 (Fig. 12.15) (Loomis, 1974). The inhibition seems to be fairly specific since

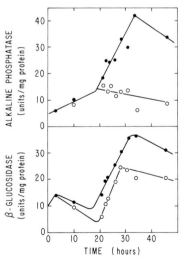

Fig. 12.15. Effect of L-cysteine on the accumulation of alkaline phosphatase and β-glucosidase. Amoebae were deposited on filters in the presence (O———O) or absence (●———●) of $10^{-2}M$ L-cysteine (Loomis, 1974).

accumulation of β-glucosidase-2 is unaffected by the compound. Gluta-thionine, another sulfhydryl reducing agent, has no effect on morpho-genesis or accumulation of alkaline phosphatase-2 when present at 10^{-2} M. At this concentration both dithiothreoitol and mercaptoethanol block morphogenesis at the aggregation stage (Gerisch, 1962c). Neither of the late enzymes accumulate in cells treated with 10^{-2} M solutions of these reducing agents. Since L-cysteine completely inhibits alkaline phosphatase-2 activity without affecting morphogenesis, it appears that the enzyme is not essential for the formation of apparently normal fruit-ing bodies. This isozyme probably functions during germination.

Ethylurethan disrupts the normal process of culmination and leads to a truncated fruiting body in which vacuolized stalk cells surround a mass of spores (Gerisch, 1961b). At other times a ball of spores is found at the base of the stalk reminiscent of the slippery stalk phenotype.

Ammonia gas suppresses culmination in *D. discoideum* and leads to simple mounds of spores and undifferentiated cells reminiscent of the *Guttulina* and *Guttulinopsis* forms (see Chapter 1) (Cohen, 1953a). When added just before culmination, a partial pressure of 0.5 mm Hg of NH_3 blocks encapsulation (Loomis, 1968).

Various salts affect culmination of *D. discoideum* (Maeda, 1970). Lithium ions in the presence of Mg^{2+} shift the proportions of stalk and spore cells in favor of stalk cells while fluoride ions have the converse effect. In the absence of Ca^{2+} ions, Na^+ ions can completely block morpho-genesis at 20 mM. Together these results implicate the ionic balance in the mechanisms of tissue proportioning and morphogenesis of *D. discoideum*.

Growth Conditions

The pattern of accumulation of several of the stage specific enzymes is modified when axenic strains are grown in broth medium rather than on bacteria (Quance and Ashworth, 1972). The specific activities of both N-acetylglucosaminidase and α-mannosidase are severalfold higher in A3 cells growing exponentially in HL-5 medium compared to A3 cells growing exponentially on bacteria (Fig. 12.16). A likely explanation for the precocious synthesis of these enzymes in axenically grown cells is that they are subject during the growth stage to some of the signals which initiate development when bacterially grown cells are starved. Cells growing in axenic broth multiply at less than half the rate of cells growing on bacteria and are probably limited for nutrients.

When Ax-2 cells are grown in broth medium lacking glucose, the

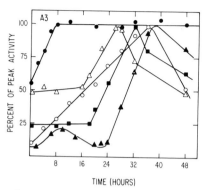

Fig. 12.16. Stage-specific enzymes in axenically grown cells of strain A3. Amoebae were harvested from HL-5 broth medium during the exponential phase of growth and deposited on filter supports. Samples were taken and assayed for: N-acetylglucosaminidase (●——●); α-mannosidase (○——○); tyrosine transaminase (△——△); alkaline phosphatase (■——■); and β-glucosidase (▲——▲).

initial concentration of glycogen is greatly reduced compared to that of cells grown in complete medium (see Chapter 8). The conditions of growth affect the peak activity of glycogen phosphorylase (Fig. 12.17) (Garrod and Ashworth, 1973). Cells with an initially low glycogen content accumulate only about half as much enzyme as do those with an initially high glycogen content. Fruiting bodies which have developed from cells with a low initial glycogen content are smaller and have a lower proportion of stalk cells (Garrod and Ashworth, 1972). It is

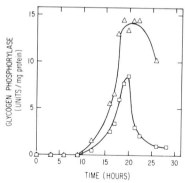

Fig. 12.17. Changes in specific activity of glycogen phosphorylase during the differentiation of amoebae of strain Ax-2 containing initially 4.9 mg glycogen/10^8 cells (△——△) or 0.3 mg glycogen/10^8 cells (□——□). [Adapted from Garrod and Ashworth (1973). Development of the cellular slime mould *Dictyostelium discoideum. Symp. Soc. Gen. Microbiol.* **23**, 407–435. With permission of Cambridge University Press.

not clear whether the effect of glycogen content on glycogen phosphorylase and tissue proportioning are related.

Genetic Modifications

Many of the modifications of morphogenesis which occur in temporal and morphological mutant strains have been described in Chapters 3 and 9. The mutations affect many of the stages through which *D. discoideum* develops and indicate alternate routes open to the organism. Temperature-sensitive mutants can also be used to delineate specific processes and the regulation of biochemical differentiation (Loomis, 1969b). Strains TS 2 and DTS 6 develop normally at the permissive temperature of 22°C and accumulate each of the stage-specific enzymes analyzed (Fig. 12.18). At the nonpermissive temperature, 27°C, cells of strain TS 2 fail to aggregate and remain as a smooth lawn of cells while

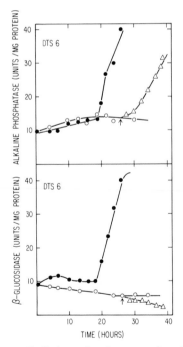

Fig. 12.18. Accumulation of alkaline phosphatase and β-glucosidase in strain DTS 6. The cells were incubated either at 22°C (●———●) or at 27°C (○———○). After 28 hours at 27°C some cells were shifted to 22°C and incubated further (△———△) (Loomis, 1974).

cells of strain DTS 6 aggregate but fail to transform into pseudoplas-modia. The early enzymes, N-acetylglucosaminidase and α-mannosidase, accumulate in these strains at 27°C but the late enzymes, β-glucosidase-2 and alkaline phosphatase-2, do not appear. Clearly a process necessary for accumulation of the late stage-specific enzymes fails to occur at the nonpermissive temperature.

The temperature-sensitive period can be delineated by allowing development to occur at 22°C until the point at which it can proceed at 27°C. For both TS 2 and DTS 6 this was found to occur after 10 hours of development at the permissive temperature. After this period they develop well at 27°C and accumulate the late enzymes normally (Loomis *et al.*, unpublished). It appears that processes occur during the first 10 hours of development that are necessary for subsequent development.

Incubation of the developmentally temperature-sensitive strains at the nonpermissive temperature does not permanently block development since on shifting cells down to 22°C after 26 hours at 27°C, development proceeds normally in both strains. The late enzymes accumulate normally in strain TS 2 after the shift down. However, in strain DTS 6 β-glucosidase-2 does not accumulate after a shift down from 27°C (Fig. 12.18). Alkaline phosphatase-2 accumulates normally under these conditions in this strain. Thus, it appears that incubation at 27°C results in a specific irreversible block to accumulation of β-glucosidase-2. This defines a branch point in the developmental program presented in Fig. 9.14. It also demonstrates that β-glucosidase-2 plays no obligatory role in culmination or formation of apparently normal fruiting bodies. The enzyme may play an essential role in germination.

Synergy

The presence of a few wild-type cells in a population of aggregateless cells will often result in differentiation of the mutant cells and formation of viable spores of both genotypes (Ennis and Sussman, 1958a). The wild-type cells seem to provide some process missing in the aggregateless strain. As discussed in Chapter 5, there are several different classes of aggregateless mutants and not every one responds to the presence of wild-type cells. However, no systematic analysis of the interactions has been made.

Attempting to demonstrate a diffusable component involved in the interaction, Sussman and Lee (1955) incubated wild-type and mutant cells on opposite sides of agar membranes only 30 μm thick. In no case did the aggregateless strains proceed through morphogenesis under these conditions. It would seem that cell–cell contact is necessary for the strains to interact.

Biochemical differentiations proceed in an essentially normal fashion in synergistic mixtures of morphological mutants and wild-type amoebae where the ratio of cell types is as great as 10 to 1 (Yanagisawa, Loomis and Sussman, 1967). For instance, UDPgalactose transferase has been shown to accumulate with normal kinetics in mixtures of strains KY-11 and NC-4 (Fig. 12.19). Moreover, 5 stage-specific enzymes have been found to accumulate to almost normal peak specific activity in mixtures of either of 2 aggregateless strains, 111 or 115, with strain NC-4 (Table 12.1). These results indicate that once induced by the presence of wild-type cells, these aggregateless strains can proceed normally through development.

Two independent morphological mutant strains, neither of which will develop alone, will often form fruiting bodies when incubated together (Sussman and Lee, 1955; Yamada et al., 1973). In one such pairing, an aggregateless strain and a strain which formed irregular mounds, formed mature fruiting bodies when incubated together and accumulated UDPgalactose polysaccharide transferase to 42% of peak specific activity (Yanagisawa et al., 1967). Alone, neither strain accumulated this enzyme. This kind of synergistic development is most frequently observed in mixtures of strains affected at different stages rather than in mixtures of strains affected at the same morphological stage (Yamada et al., 1973). However, there is little resolution in staging on the basis of visual morphology and exceptions to this rule have been found. When the biochemical basis for the interactions are better understood, the pattern of synergistic competence may delineate significant steps in development.

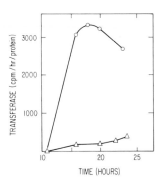

Fig. 12.19. Accumulation of UDPgalactose polysaccharide transferase in strain KY-11. Aggregateless strain KY-11 was incubated alone (△——△) or in association with wild-type cells present at 10% the number of mutant cells (○——○) (Yanagisawa et al., 1967).

TABLE 12.1
SYNERGY OF AGGREGATELESS STRAINS

	Percent of maximal activity[a]				
Enzyme	NC-4 alone	111 alone	111+NC-4 synergistically	115 alone	115+NC-4 synergistically
Acetylglucosaminidase	100	105	100	130	93
Threonine deaminase-2	100	2	74	0.5	69
Tyrosine transaminase	100	33	115	36	111
Alkaline phosphatase-2	100	0.5	96	0	71
β-Glucosidase-2	100	3	63	6	80

[a] Strains NC-4 and aggregateless mutant strains 111 and 115 were incubated on filter supports either alone or in synergistic pairs. Samples were collected in the presence of pyridoxal phosphate at intervals over 28 hours after the initiation of development. Stage-specific enzymes were assayed in frozen and thawed samples after sonification in at least 2 separate experiments. Peak activity in 2 separate samples were compared to peak activity in wild-type cells. Peak activity in wild-type (NC-4) cells: N-acetylglucosamindase 230 units/mg; threonine deaminase-2, 6.8 units/mg; tyrosine transaminase, 44 units/mg; alkaline phosphatase-2, 32 units/mg; β-glucosidase-2, 30 units/mg.

Clearly, development of *D. discoideum* is sensitive to a variety of modifications from the environment. Many of these remain as enigmas which challenge future elucidation. Others appear to have direct bearing on our present concepts. One of the most interesting sets of phenomena is the biochemical response to dissociation. The results appear to indicate that cell–cell interactions play an essential role in integration of differentiation, critical mass determination and tissue proportioning. These are discussed further in the following chapter.

Chapter 13

Tissue Proportioning

Size Invariance

Pseudoplasmodia containing as many as 10^5 or as few as a hundred cells form fruiting bodies of the same general shape and apparent proportions. Somehow the number of cells which differentiate into spores is regulated with respect to the total number of cells in a pseudoplasmodium. Bonner and Slifkin (1949) measured the volume of stalks and sori in fruiting bodies which varied 30-fold in total volume and found that the ratio of the volumes of stalks and sori varied only 2-fold (Fig. 13.1). Large fruiting bodies had a somewhat higher proportion of spores but there was considerable variation among individual fruiting bodies of the same size. Considering that the analysis included fruiting bodies of greatly differing size, it appears that mechanisms function in *D. discoideum* which insure that the proportion of cells that differentiate into spores is essentially invariant with respect to the total number of cells. The volume of the spore mass was found to be about twice that of

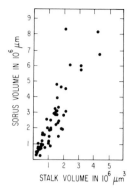

Fig. 13.1. Proportions of stalk and spore-mass volume in fruiting bodies of varying size. Development proceeded at 17°C in the dark. Volumes were calculated from camera lucida drawings (Bonner and Slifkin, 1949).

the stalk in both large and small fruiting bodies. The actual ratio of spore cells to stalk cells is somewhat greater than 2 to 1 since stalk cells increase severalfold in volume when they vacuolate. A more accurate determination of the proportion of cells differentiating into spore or stalk cells was made by measuring the dry weight of isolated spores and stalks from fruiting bodies of various sizes (Bonner, 1952; Farnsworth, 1974). The dry weight of spores from fruiting bodies which had developed at 22°C was found to be 83 ± 1% of the total.

Mechanical Analysis

It has been considered that the number of cells which differentiate into stalk cells is determined purely by the mechanics and timing of the process of culmination. Farnsworth (1973b) pointed out that the size of the anterior tip is determined by the overall size of the pseudo-plasmodium and that the tip size determines the diameter of the stalk sheath. Thus, large pseudoplasmodia will initiate large stalks while small pseudoplasmodia will initiate small stalks. As the mass of cells rises on the elongating stalk, undifferentiated cells are drawn into the stalk sheath where they are induced to vacuolate. However, spore cells do not enter the stalk sheath. When all cells outside the stalk have differentiated into spores no further increase in stalk can occur. This point of view considers the proportions in a fruiting body to be determined only by a race between encapsulation and entry into the stalk. In this model no prior cellular specialization is required and all cells are thought to be able to form either spores or stalks. However, as discussed in Chapter

7, there appear to be certain differentiations which occur only in cells in the posterior of pseudoplasmodia. At present it is not known if these biochemical differentiations direct the cells toward specific pathways of cytodifferentiation.

Differentiations of Prespore and Prestalk Cells

By observing vitally stained cells Raper (1940b) found that those in the anterior quarter of pseudoplasmodia give rise to stalk cells while those in the posterior form spores. Thus, if there are spore-specific differentiations, they would be expected to be found in the posterior cells. The prespore vesicles which are found predominantly in cells of the posterior portions of pseudoplasmodia do appear to be related to spore formation since they disappear at the same time as the cells encapsulate (Hohl and Hamamoto, 1969b; Gregg and Badman, 1970). It has been reported that cells in the anterior 10–15% are devoid of these organelles (Muller and Hohl, 1973). The vesicles contain a polysaccharide antigen which is ultimately secreted in sori (Ikeda and Takeuchi, 1971; Takeuchi, 1972). Moreover, they may contain the enzyme, UDPgalactose transferase, which is also localized exclusively in the posterior parts of pseudoplasmodia (Newell *et al.*, 1969). Although it cannot be conclusively shown that accumulation of these vesicles is essential for spore differentiation, it seems likely that this is the case.

One of Raper's (1940b) early observations suggests that anterior cells of migrating pseudoplasmodia have not undergone those differentiations necessary for immediate spore formation. When he isolated the anterior 10% of a pseudoplasmodium and induced culmination, he found that almost all the cells formed stalk cells and very few formed spores. This result was confirmed by Bonner and Slifkin (1949). Thus, the lack of prespore vesicles in anterior cells correlates with the inability of these cells to rapidly differentiate into spores.

Posterior portions of bisected pseudoplasmodia were found to form both spore and stalk cells upon culmination (Raper, 1940b). It would appear that both anterior and posterior cells can rapidly undergo differentiations necessary for stalk formation. This conclusion was further supported by the observations of Farnsworth (1973b) that cells enclosed in a nitrocellulose tube would form stalk cells whether they came from the front or the back of pseudoplasmodia.

On the other hand, there have been various reports of localization of cytological structures and biochemical activities unique to anterior cells suggesting that these cells may be specialized for stalk differentiation. Miller *et al.* (1969) reported a higher concentration of sheets of double membranes in anterior cells. However, this has not been found in several other laboratories (Muller, 1972; Maeda *et al.*, 1973; Farnsworth, personal communication). No other cytological structures have been seen to occur exclusively in the prestalk cells. Using histochemical techniques, Krivanek and Krivanek (1958) and Mine and Takeuchi (1967) have reported that several enzymes, including alkaline phosphatase and succinoxidase, are localized in anterior cells. However, when enzymatic assays were performed on cell-free extracts of isolated anterior or posterior cells no significant differences were found in these specific activities (Farnsworth and Loomis, 1974). As is discussed below, the histochemical results can be explained by the pattern of diffusion of the substrates or chromogen into pseudoplasmodia. Thus, there is presently no evidence for any biochemical properties of stalk cells which are not common to prespore cells. The following discussion, therefore, considers only the problem of restriction of certain differentiations to cells in the posterior of pseudoplasmodia.

Cell Sorting and Determination

Several workers have suggested that the proportion of spores and stalk cells are determined prior to aggregation (Bonner, 1957, 1959a; Takeuchi, 1969; Francis and O'Day, 1971; Muller and Hohl, 1973). Bonner (1957, 1959a) has argued that certain cells in a growing population have a tendency to form spores while others tend to form stalk cells. He proposes that cells determined to form spores migrate more slowly within pseudoplasmodia than those which are determined to form stalk and therefore take up a posterior position. He suggests that considerable cell sorting occurs with pseudoplasmodia. However, as is discussed in Chapter 6, present evidence indicates that little or no cell mixing takes place in pseudoplasmodia. Thus, if cell sorting of predisposed cells occurs, it must take place at an earlier stage.

Takeuchi (1969) studied the tendency of cells dissociated from anterior and posterior fragments of pseudoplasmodia to take up their original position when mixed and allowed to reaggregate. He grew cells with bacteria in the presence of ^{3}H-thymidine so as to label DNA. Labeled cells from the posterior of migrating pseudoplasmodia were mixed with unlabeled cells from the anterior of other pseudoplasmodia and allowed to reaggregate. The distribution of label in the resulting pseudoplasmodia was determined autoradiographically, and the posterior regions were found to give rise to 3–5 times as many grains as the anterior regions.

Takeuchi suggested that the cells had sorted out during aggregation and returned to their previous axial positions, perhaps as a consequence of the differential cohesiveness of the 2 cell types. However, it is not clear that the label was restricted to DNA and gave an accurate indication of the position of the labeled cells. The distribution may have resulted from other effects. For instance, when cells were labeled with ^{3}H-uridine, a gradient in counts was observed along the length of pseudoplasmodia even when a homogeneous population of cells was allowed to aggregate (Takeuchi, 1969). Thus, the anterior–posterior gradient may have formed by mechanisms other than cell sorting, perhaps as a consequence of polarized migration. Francis and O'Day (1971) have shown that more than 20% of the label in pseudoplasmodia formed from cells grown on ^{3}H-thymidine is acid soluble. These low molecular weight compounds may diffuse within pseudoplasmodia or adhere to the sheath as it passes over the cells. This uncertainty in the localization of the label, as well as the low number of counts present in labeled pseudoplasmodia (20–30 counts above background), makes the interpretation of these results difficult and leaves the possibility of differential cohesion and subsequent sorting out of cells from anterior and posterior positions an open question.

Regulation of Tissue Proportions

The classic experiments of Raper (1940b) give direct evidence that cells are not irreversibly committed to spore differentiation at any stage prior to terminal cytodifferentiation. He found that posterior pseudoplasmodial cells, which would all differentiate into spores if left *in situ*, form both spore and stalk cells when posterior fragments were isolated and induced to culminate immediately. Although isolated anterior fragments form only stalk cells when induced to culminate immediately after being removed from pseudoplasmodia, they form both spore and stalk cells if allowed to migrate for 4 hours or more (Raper, 1940b). With progressively longer periods of migration, the ratio of spores to stalk cells approaches that of normal fruiting bodies. These results suggest that posterior cells can rapidly undergo stalk differentiations but that a period of migration is essential for the cells in the back of isolated anterior fragments to undergo differentiations necessary for spore formation.

Initially all cells in the anterior fragments are devoid of prespore vesicles but they accumulate these organelles following isolation (Gregg and Badman, 1970; Sakai and Takeuchi, 1971). Gregg and Badman

(1970) reported that prespore vesicles could be observed in anterior fragments within an hour after isolation. However, they appear to have isolated fragments which were too large a proportion of the pseudoplasmodia since normally proportioned fruiting bodies were formed within a few hours and in absence of migration. Sakai and Takeuchi (1971), on the other hand, reported that prespore vesicles accumulated in isolated anterior fragments only after 5 hours of migration. Thus, at the cytological level it can be seen that cells within pseudoplasmodia are able to regulate in accord with their new position when the size of the field is changed, but that this process takes several hours.

Positional Information

Wolpert (1971) has considered the problem of pattern formation and tissue proportioning and has developed concepts which can be used in analyzing a variety of regulative systems. Wolpert suggests that each cell might receive, from some signal, a positional value which specifies its axial position. The cells may then interpret this positional value by turning on the genes appropriate for differentiation of cells in such a position. Those cells, which have their position specified with respect to the same set of points, are defined as a field which for *D. discoideum* is an individual pseudoplasmodium.

Several signaling mechanisms have been proposed which might give cells information about position in pseudoplasmodia of *D. discoideum.* An organizing region located at one end of the field which gives off one or more periodic signals could provide the cells with positional values if they were able to interpret the signal by temporal integration. Since aggregation is known to involve a pulsatile signal in *D. discoideum,* it has been considered that this signaling process may function to give positional information in pseudoplasmodia (Robertson and Cohen, 1972). However, a related species, *D. minutum,* which lacks pulsatile aggregation, is nevertheless able to establish a normal pattern. Likewise, mutants of *D. discoideum,* in which the pattern of pulsation is modified, form well-proportioned fruiting bodies (Gerisch, 1968, 1971; Durston, 1974). Thus, periodic signaling does not seem to be essential for the normal pattern of differentiation in this system.

Crick (1970) has suggested that the involvement of diffusable compounds acting as morphogens should be seriously considered because of the simplicity of the mechanisms involved. He considered the case in which a source occurs at one end of a field and a sink at the other. If the concentration of the morphogen is held constant at the ends, a monotonic gradient will be established across the field. The time it would take to establish such a gradient over a field the size of pseudo-plasmodia (about 1 mm) has been calculated for a substance of 500 daltons (Crick, 1970). Taking into account a variety of variables, the gradient should be established within a few hours. This is within the period of time required for prespore vesicles to accumulate in isolated anterior fragments (Sakai and Takeuchi, 1971).

In pseudoplasmodia of *D. discoideum* the pattern consists of only 2 cell types: those with prespore vesicles and those lacking them. A morphogen eliciting specific differentiations in prespore cells could be generated at the posterior end and removed at the anterior. Alternatively an inhibitor of specific differentiations could be produced at the anterior end and removed at the posterior. As discussed previously, there is no conclusive evidence that biochemical or cytological differentiations are localized to the anterior cells and so positional effects on differentiations of prestalk cells are not considered in this discussion.

In generalized models the specialized nature of the source and the sink are thought to result from their position at the boundaries of the field. When the system is perturbed, such as by removal of either end of the field, the gradient of morphogen is reestablished by regions which now find themselves at the boundaries of the new field and take over the function of source or sink. These models can account in general terms for the regulatory properties and size invariance observed in the development of *D. discoideum*, but do not explain the specialized properties of the boundaries.

Cell Contact Model

McMahon (1973) has presented a detailed model which suggests that there are "contact-sensing" molecules on the surface of cells that regulate the internal concentration of a substance, "A," which provides positional value. He proposes that the contact-sensing molecules are activated by interaction with complementary molecules on adjacent cells. Two kinds of contact-sensing molecules are postulated: one increases the concentration of "A," the other decreases it. The 2 types are localized in separate areas of the cell surface, one at the front of the cell, the other at the rear. This localization is postulated to be a constant feature of the cells.

Using these postulates McMahon has shown that with appropriate parameters the model can generate a 10-fold difference in concentration of "A" from the front to the back of a pseudoplasmodium of *D. dis-coideum*. The distinguishing characteristic of this model is that pattern is dependent on polarity and the polarity of the system is generated as a consequence of the affinity of the complementary surface molecules.

McMahon suggests that cAMP may function as the morphogen "A" in *Dictyostelium* and other systems. By inserting parameters of the appropriate *Dictyostelium* enzymes, such as adenyl cyclase and phosphodiesterase, into equations derived from his postulates he has shown that the model predicts that a sharp increase in cAMP could be generated in the anterior third of a pseudoplasmodium within 3 hours. This difference in cAMP could be interpreted by the cells to result in tissue proportioning: anterior cells could be inhibited from undergoing specific differentiations. Unless other specific pseudoplasmodial characteristics are assumed to be required, this model would predict that isolated cells would differentiate into spores since they would be in a condition of low cAMP concentration. This, however, has not been observed. Although the model has several attractive features, recent evidence does not support the prediction of a gradient in cAMP along the axis of pseudoplasmodia. Direct analyses of isolated anterior and posterior portions have shown little difference in cAMP concentration (Garrod and Malkinson, 1973; Brenner, personal communication; Farnsworth, unpublished).

Diffusional Barrier Model

I have proposed that a gradient in morphogen could be established if diffusion of certain molecules to the environment was limited except at the front of pseudoplasmodia (Loomis, 1972). The concentration of a morphogen would then be highest at the back if all cells produced and released the compound. The model suggests that cells respond to concentrations of a morphogen over a threshold value by undergoing differentiations necessary for spore formation. I proposed that the surface sheath acts as a barrier to diffusion of morphogen out of a pseudoplasmodium except at the anterior end.

Direct observation of the thickness of the surface sheath after fixation shows that it is thinnest over the anterior tip and the thickness increases with increasing distance back from the tip (Fig. 6.1) (Farnsworth and Loomis, in press). Moreover, the sheath is extensible only at the tip and thus appears to be qualitatively different in this region (Francis, 1962). Sheath newly deposited at the anterior tip may be less of a barrier to diffusion.

Since we have no idea of the nature of the hypothetical morphogen in this or any other system, we cannot directly test the effect of the surface sheath on diffusion of this molecule. However, we have been

able to measure diffusion of various other small molecules into pseudoplasmodia (Farnsworth and Loomis, 1974). The entry of the dye, nitro-
blue tetrazolium (NBT), can be followed since this compound forms
a dark blue precipitate when reduced by succinoxidase activity. The
specific activity of this enzyme system measured *in vitro* with NBT
is identical in anterior and posterior cells (Farnsworth and Loomis,
1974). However, when intact pseudoplasmodia are incubated in the
presence of succinate and NBT, the anterior portions stain far more
rapidly than do posterior portions (Fig. 13.2); incubation with NBT
prior to addition of succinate results in rapid uniform staining. Thus,
the anterior staining upon simultaneous addition of succinate and NBT
seems to be a consequence of an axial inhomogeneity in a barrier to
diffusion of NBT into pseudoplasmodia.

This barrier to diffusion appears to be the surface sheath since, when
it is punctured in the posterior region of pseudoplasmodia, the underlying
cells are stained as rapidly as anterior cells when incubated with succinate and NBT (Farnsworth and Loomis, 1974). We would expect
the sheath to be an equally effective barrier to diffusion out of pseudoplasmodia of similar sized compounds released by the cells. It is clear
that a gradient in diffusable compounds must exist within pseudoplasmodia but it is still unclear whether the cells use this positional information to establish tissue proportioning.

The model has an important feature which distinguishes it from the
general models considered by Wolpert (1971) and Crick (1970): no
localized source is invoked and a sink is maintained as a consequence
of cell movement. The model also suggests that polarity results from

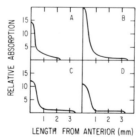

Fig. 13.2. Diffusion into pseudoplasmodia. Pseudoplasmodia of varying length were
incubated in a solution containing 2 m*M* succinate, 1 mg/ml nitrotetrazoalium blue,
and phosphate buffer pH 7.5. After various periods of time the pseudoplasmodia
were washed free of the reagents, pressed onto a slide with a cover slip, and the
absorbance measured on a Joyce-Loebl microspectrophotometer using a red filter.
The spectrophotometer was blanked against an unstained pseudoplasmodia. Incubation times were (A) 18 minutes, (B) 20 minutes, (C) 37 minutes, and (D) 37
minutes (Farnsworth and Loomis, 1974).

the mechanism of pseudoplasmodial formation and migration. Moreover, the model can be adapted to account for regulation and pattern formation in a variety of other systems.

Structural Model

Ashworth (1971) has proposed that cells in pseudoplasmodia of *D. discoideum* gain positional information from the chemical structure of the sheath itself. He suggested that the sheath could be chemically modified as the cells move over it, perhaps by addition or removal of specific carbohydrate side chains by cell-bound enzymes. Since the sheath surrounding posterior cells has passed over all preceding cells it would be subject to greater modification than sheath surrounding anterior cells. Thus, a gradient in chemical modifications of the sheath would occur down the axis which could be interpreted by the cells to control the pattern of differentiation. The interaction of the cell surface with the sheath modifications would have to be transmitted to underlying cells by some mechanism involving direct contact since all posterior cells, not just those at the surface, undergo the differentiations necessary for spore formation. However, cells must communicate the positional information only toward the center of the pseudoplasmodium and not along the length. The model does not suggest how cells communicate their response to the sheath in a radial but not an axial fashion.

The degree of modification of the sheath will depend directly on the length of the pseudoplasmodium and the rate of migration. Since the axial ratio of both large and small pseudoplasmodia is constant, the rate of migration appears to depend on the length of the pseudoplasmodium. Therefore, this model suggests a constant degree of modification of the sheath in both large and small pseudoplasmodia. The model can thus account for size invariance of tissue proportioning.

It is clear from the variety of models which have been presented that much more information on the mechanism of tissue proportioning in *Dictyostelium* is needed. The models serve to focus our attention on the crucial aspects of positional information in this system and direct further studies toward potentially rewarding approaches. When the actual mechanism for the generation and interpretation of positional information is determined in this relatively simple system, it will be interesting to determine whether similar mechanisms function in higher organisms.

Bibliography

Aldrich, H. C., and Gregg, J. H. (1973). Unit membrane structural changes following cell association in *Dictyostelium*. *Exp. Cell Res.* **81**, 407–412.

Anderson, J. S. (1973). A technique for clonal analysis of developmental mutants of the cellular slime mold *Dictyostelium discoideum*. *Genetics* **2**, 7–11.

Anderson, J. S., Fennell, D. I., and K. B. Raper (1968). *Dictyostelium deminutivum*, a new cellular slime mold. *Mycologia* **40**, 49–64.

Ashworth, J. M. (1966). Studies on the ribosomes of the cellular slime molds. *Biochim. Biophys. Acta* **129**, 211–213.

Ashworth, J. M. (1968). The life and times of the slime mould. *New Sci. June*, 629–631.

Ashworth, J. M. (1971a). Cell development in the cellular slime mould *Dictyostelium discoideum*. *Symp. Soc. Exp. Biol.* **25**, 27–49.

Ashworth, J. M. (1971b). "Control Mechanisms of Growth and Differentiation" (D. D. Davies, and M. Balls, eds.), Vol. 25, pp. 27–49.

Ashworth, J. M., and Quance, J. (1972). Enzyme synthesis in myxamoebae of the cellular slime mould *Dictyostelium discoideum* during growth in axenic culture. *Biochem. J.* **126**, 601–608.

Ashworth, J. M., and Sackin, M. J. (1969). Role of aneuploid cells in cell differentiation in the cellular slime mould *Dictyostelium discoideum*. *Nature (London)* **224**, 817–818.

Ashworth, J. M., and Sussman, M. (1967). The appearance and disappearance of uridine diphosphate glucose pyrophosphorylase activity during differentiation of the cellular slime mold *Dictyostelium discoideum*. *J. Biol. Chem.* **242**, 1696–1700.

Ashworth, J. M., and Watts, D. J. (1970). Metabolism of the cellular slime mould *Dictyostelium discoideum* grown in axenic culture. *Biochem. J.* **119**, 175–182.

Ashworth, J. M., and Weiner, E. (1972). The lysosomes of the cellular slime mould *Dictyostelium discoideum*. *In* "Lysosomes in Biology and Pathology" (J. T. Dingle, ed.), Vol. 3, pp. 36–46. North-Holland Publ., Amsterdam.

Ashworth, J. M., Duncan, D., and Rowe, A. J. (1969). Changes in fine structure during cell differentiation of the cellular slime mold, *Dictyostelium discoideum*. *Exp. Cell. Res.* 58, 73–78.

Bacon, C. W., Sussman, A. S., and Paul, A. G. (1973). Identification of a self-inhibitor from spores of *Dictyostelium discoideum*. *J. Bacteriol.* 113, 1061–1063.

Barkley, D. S. (1969). Adenosine-3′,5′-phosphate: identification as acrasin in a species of cellular slime mold. *Science* 165, 1133–1134.

Bauer, R., Rath, M., and Risse, H. (1971). The biosynthesis of glycoproteins during the development of *Dictyostelium discoideum*. The transfer of D-mannose in vegetative and aggregated cells. *Eur. J. Biochem.* 21, 179–190.

Baumann, P. (1969). Glucokinase of *Dictyostelium discoideum*. *Biochemistry* 8, 5011–5015.

Baumann, P., and Wright, B. E. (1968). The phosphofructokinase of *Dictyostelium discoideum*. *Biochemistry* 7, 3653–3661.

Baumann, P., and Wright, B. E. (1969). The fructose 1,6-diphosphatase of *Dictyostelium di coideum*. *Biochemistry* 8, 1655–1659.

Beug, H., and Gerisch, G. (1969). Univalente fragmente von antikörpern zur analyse von zellmembran-funktionen. *Naturwissenschaften* 56, 374.

Beug, H., and Gerisch, G. (1972). A micromethod for routine measurement of cell agglutination and dissociation. *J. Immunol. Methods* 2, 49–57.

Beug, H., Gerisch, G., Kempf, S., Riedel, V., and Cremer, G. (1970). Specific inhibition of cell contact formation in *Dictyostelium* by univalent antibodies. *Exp. Cell. Res.* 63, 147–158.

Beug, H., Katz, F. E., and Gerisch, G. (1973a). Dynamics of antigenic membrane sites relating to cell aggregation in *Dictyostelium discoideum*. *J. Cell Biol.* 56, 647–658.

Beug, H., Katz, F. E., Stein, A., and Gerisch, G. (1973b). Quantitation of membrane sites in aggregating *Dictyostelium* cells by use of tritiated univalent antibody. *Proc. Nat. Acad. Sci. U.S.* 70, 3150–3154.

Blaauw, A. (1909). Die perzeption des lichtes. *Rec. Trav. Bot. Neer.* 5, 209–372.

Blaskovics, J. A., and Raper, K. B. (1957). Encystment stages of *Dictyostelium*. *Biol. Bull.* 113, 58–88.

Bonner, J. T. (1944). A descriptive study of the development of the slime mold *Dictyostelium discoideum*. *Amer. J. Bot.* 31, 175–182.

Bonner, J. T. (1947). Evidence for the formation of cell aggregates by chemotaxis in the development of the slime mold *Dictyostelium discoideum J. Exp. Zool.* 106, 1–26.

Bonner, J. T. (1949). The demonstration of acrasin in the later stages of the development of the slime mold *Dictyostelium discoideum*. *J. Exp. Zool.* 110, 259–271.

Bonner, J. T. (1950). Observations on polarity in the slime mold *Dictyostelium discoideum*. *Biol. Bull.* 99, 143–151.

Bonner, J. T. (1952).The pattern of differentiation in amoeboid slime molds. *Amer. Natur.* 86, 79–89.

Bonner, J. T. (1957). A theory of the control of differentiation in the cellular slime molds. *Quart. Rev. Biol.* 32, 232–246.

Bonner, J. T. (1959a). Evidence for the sorting out of cells in the development of the cellular slime molds. *Proc. Nat. Acad. Sci. U.S.* 45, 379–384.

Bonner, J. T. (1959b). "The Cellular Slime Molds," 1st ed. Princeton Univ. Press, Princeton, New Jersey.

Bonner, J. T. (1959c). Differentiation of social amoebae. *Sci. Amer.* **201**, 152–162.

Bonner, J. T. (1960). Development in the cellular slime molds: the role of cell division, cell size, and cell number. *In* "Developing Cell Systems and Their Control" (D. Rudnick, ed.), 18th Growth Symposium, pp. 3–20. Ronald Press, New York.

Bonner, J. T. (1963a). How slime molds communicate. *Sci. Amer.* **209**, 84–93.

Bonner, J. T. (1963b). Epigenetic development in the cellular slime molds. *Symp. Soc. Exp. Biol.* **17**, 341–358.

Bonner, J. T. (1965). Physiology of development in cellular slime molds. (Acrasiales). *In* "Encyclopedia of Plant Physiology" (W. Ruhland, ed.), Vol. 15, Part 1, pp. 612–640.

Bonner, J. T. (1967). "The Cellular Slime Molds," 2nd ed., 205 pp. Princeton Univ. Press, Princeton, New Jersey.

Bonner, J. T. (1969). Hormones in social amoebae and mammals. *Sci. Amer.* **220**, 78–91.

Bonner, J. T. (1970). Induction of stalk cell differentiation by cyclic AMP in the cellular slime mold *Dictyostelium discoideum*. *Proc. Nat. Acad. Sci. U.S.* **65**, 110–113.

Bonner, J. T. (1971). Aggregation and differentiation in the cellular slime molds. *Annu. Rev. Microbiol.* **25**, 75–92.

Bonner, J. T., and Adams, M. S. (1958). Cell mixtures of different species and strains of cellular slime moulds. *J. Embryol. Exp. Morphol.* **6**, 346–356.

Bonner, J. T., and Dodd, M. R. (1962a). Aggregation territories in the cellular slime molds. *Biol. Bull.* **122**, 13–24.

Bonner, J. T., and Dodd, M. R. (1962b). Evidence for gas-induced orientation in the cellular slime molds. *Develop. Biol.* **5**, 344–361.

Bonner, J. T., and Eldredge, D. Jr. (1945). A note on the rate of morphogenetic movement in the slime mold, *Dictyostelium discoideum*. *Growth* **9**, 287–297.

Bonner, J. T., and Frascella, E. B. (1952). Mitotic activity in relation to differentiation in the slime mold *Dictyostelium discoideum*. *J. Exp. Zool.* **121**, 561–571.

Bonner, J. T., and Frascella, E. B. (1953). Variations in cell size during the development of the slime mold, *Dictyostelium discoideum*. *Biol. Bull.* **104**, 297–300.

Bonner, J. T., and Hoffman, M. E. (1963). Evidence for a substance responsible for the spacing pattern of aggregation and fruiting in the cellular slime molds. *J. Embryol. Exp. Morphol.* **11**, 571–589.

Bonner, J. T., and Shaw, M. J. (1957). The role of humidity in the differentiation of the cellular slime molds. *J. Cell. Comp. Physiol.* **50**, 145–154.

Bonner, J. T., and Slifkin, M. K. (1949). A study of the control of differentiation: the proportions of stalk and spore cells in the slime mold *Dictyostelium discoideum*. *Amer. J. Bot.* **36**, 727–734.

Bonner, J. T., and Whitfield, F. E. (1965). The relation of sorocarp size to phototaxis in the cellular slime mold *Dictyostelium prupureum*. *Biol. Bull.* **128**, 51–57.

Bonner, J. T., Clarke, W. W., Jr., Neely, C. L., Jr., and Slifkin, M. K. (1950). The orientation to light and the extremely sensitive orientation to temperature gradients in the slime mold *Dictyostelium discoideum*. *J. Cell. Comp. Physiol.* **36**, 149–158.

Bonner, J. T., Koontz, P. G., Jr., and Paton, D. (1953). Size in relation to the rate of migration in the slime mold *Dictyostelium discoideum*. *Mycologia* **45**, 235–240.

Bonner, J. T., Chiquoine, A. D., and Kolderie, M. Q. (1955). A histochemical study of differentiation in the cellular slime molds. *J. Exp. Zool.* 130, 133–158.

Bonner, J. T., Kelso, A. P., and Gillmor, R. G. (1966). A new approach to the problem of aggregation in the cellular slime molds. *Biol. Bull.* 130, 28–42.

Bonner, J. T., Barkley, D. S., Hall, E. M., Konijn, T. M., Mason, J. W., O'Keefe, G., III, and Wolfe, P. B. (1969). Acrasin, Acrasinase, and the sensitivity to acrasin in *Dictyostelium discoideum. Develop. Biol.* 20, 72–87.

Bonner, J. T., Hall, E. M., Sachsenmaier, W., and Walker, B. K. (1970). Evidence for a second chemotactic system in the cellular slime mold, *Dictyostelium discoideum. J. Bacteriol.* 102, 682–687.

Bonner, J. T., Sieja, T. W., and Hall, E. M. (1971). Further evidence for the sorting out of cells in the differentiation of the cellular slime could *Dictyostelium discoideum. J. Embryol. Exp. Morphol.* 25, 457–465.

Bonner, J. T., Hall, E. M., Noller, S., Oleson, F., and Roberts, A. (1973). Synthesis of cyclic AMP and phosphodiesterase in various species of cellular slime molds and its bearing on chemotaxis and differentiation. *Develop. Biol.* R, 402–409.

Born, G. V. R., and Garrod, D. (1968). Photometric demonstration of aggregation of slime mould cells showing effects of temperature and ionic strength. *Nature (London)* 220, 616–618.

Bradley, S. G., and Sussman, M. (1952). Growth of ameboid slime molds in one-membered cultures. *Arch. Biochem. Biophys.* 39, 462–463.

Bradley, S. G., and Sussman, M. (1954). Physiology of the aggregation stage in the development of the cellular slime molds (abstract). *Bacteriol. Proc.* 32.

Bradley, S. G., Sussman, M., and Ennis, H. L. (1956). Environmental factors affecting the aggregation of the cellular slime mold, *Dictyostelium discoideum. J. Protozool.* 3, 33–38.

Braun, V., Hautke, K., Wolf, H., and Gerisch, G. (1972). Degradation of the murein lipoprotein complex of *Escherichia coli* cell walls by *Dictyostelium* amoebae. *Eur. J. Biochem.* 27, 116–125.

Brefeld, O. (1869). *Dictyostelium mucoroides.* Ein neuer organismus aus der verwandtschaft der myxomyceten. *Abhandl. Senckenberg. Naturforsch. Ges.* 7, 85–107.

Brefeld, O. (1884). *Polysphondylium violaceum* und *Dictyostelium mucoroides* nebst bemerkungen zur systematik der schleimpilze. *Untersuchungen aus dem Gesammtgebiet der Mykol.* 6, 1–34.

Brühmüller, M., and Wright, B. E. (1963). Glutamate oxidation in the differentiating slime mold. II. Studies *in vitro. Biochim. Biophys. Acta* 71, 50–57.

Cavender, J. C. (1972). Cellular slime molds in forest soils of Eastern Canada. *Can. J. Bot.* 50, 1497–1501.

Cavender, J. C., and Raper, K. B. (1965a). The Acrasieae in nature. I. Isolation. *Amer. J. Bot.* 52, 294–296.

Cavender, J. C., and Raper, K. B. (1965b). The Acrasieae in nature. II. Forest soil as a primary habitat. *Amer. J. Bot.* 52, 297–302.

Cavender, J. C., and Raper, K. B. (1965c). The Acrasieae in nature. III. Occurrence and distribution in forests of Eastern North America. *Amer. J. Bot.* 52, 302–308.

Cavender, J. C., and Raper, K. B. (1968). The occurrence and distribution of Acrasieae in forests of subtropical and tropical America. *Amer. J. Bot.* 55, 504–513.

Ceccarini, C. (1966). Trehalase from *Dictyostelium discoideum:* Purification and properties. *Science* 151, 454–456.

Ceccarini, C. (1967). The biochemical relationship between trehalase and trehalose during growth and differentiation in the cellular slime mold *Dictyostelium discoideum*. *Biochim. Biophys. Acta* **148**, 114–124.

Ceccarini, C., and Filosa, M. (1965). Carbohydrate content during development of the slime mold, *Dictyostelium discoideum*. *J. Cell. Comp. Physiol.* **66**, 135–140.

Ceccarini, C., and Cohen, A. (1967). Germination inhibitor from the cellular slime mold *Dictyostelium discoideum*. *Nature (London)* **214**, 1345–1346.

Ceccarini, C., and Maggio, R. (1968). Studies on the ribosomes from the cellular slime molds *Dictyostelium discoideum* and *Dictyostelium purpureum*. *Biochim. Biophys. Acta* **166**, 134–141.

Ceccarini, C., Campo, M. S., and Andronico, F. (1970). Ribosomal subunit exchange in *Dictyostelium purpureum*. *J. Cell Biol.* **46**, 428–430.

Chang, Y. Y. (1968a). Cyclic 3′,5′-adenosine monophosphate phosphodiesterase produced by the slime mold *Dictyostelium discoideum*. *Science* **160**, 57–59.

Chang, Y. Y. (1968b). Cyclic 3′5′AMP diesterase produced by the slime mold *Dictyostelium discoideum*. *Science* **161**, 57–69.

Chassy, B. M. (1972). Cyclic nucleotide phosphodiesterase in *Dictyostelium discoideum:* Interconversion of two enzyme forms. *Science* **175**, 1016–1018.

Chi, Y. Y., and Francis, D. (1971). Cyclic AMP and calcium exchange in a cellular slime mold. *J. Cell. Physiol.* **77**, 169–173.

Clarke, M., and Spudich, J. (1974). Biochemical and structural studies of cell movement. *J. Mol. Biol.* **86**, 209–222.

Clark, M. A., Francis, D., and Eisenberg, R. (1973). Mating types in cellular slime molds. *Biochem. Biophys. Res. Commun.* **52**, 672–678.

Clegg, J. S., and Filosa, M. F. (1961). Trehalose in the cellular slime mould *Dictyostelium mucoroides*. *Nature (London)* **192**, 1077–1078.

Cleland, S. V. (1969). Gluconeogenesis and glycolysis in *Dictyostelium discoideum*. Ph.D. Thesis, Northwestern Univ., Evanston, Illinois.

Cleland, S. V., and Coe, E. L. (1968). Activities of glycolytic enzymes during the early stages of differentiation in the cellular slime mold *Dictyostelium discoideum*. *Biochim. Biophys. Acta* **156**, 44–50.

Cleland, S. V., and Coe, E. L. (1969). Conversion of aspartic acid to glucose during culmination in *Dictyostelium discoideum*. *Biochim. Biophys. Acta* **192**, 446–454.

Cocucci, S., and Sussman, M. (1970). RNA in cytoplasmic and nuclear fractions of cellular slime mold amebas. *J. Cell. Biol.* **45**, 399–407.

Coemans, E. (1863). Recherches sur le polymorphisme et les différents appareils de reproduction chez les mucorinées. *Bull. Acad. Roy. Sci. Lett. Beaux Arts Belg* Ser. 2ᵉ, **16**, 177–188.

Cohen, A. L. (1953a). The effect of ammonia on morphogenesis in the Acrasieae. *Proc. Nat. Acad. Sci. U.S.* **39**, 68–74.

Cohen, A. L. (1953b). The isolation and culture of opsimorphic organisms. I. Occurrence and isolation of opsimorphic organisms from soil and culture of Acrasieae on a standard medium. *Ann. N.Y. Acad. Sci.* **56**, 938–943.

Cohen, A. L. (1965). Slime molds. *Encycl. Britannica* **20**, 797–798.

Cohen, A. L., and Ceccarini, C. (1967). Inhibition of spore germination in the cellular slime mold *Dictyostelium discoideum*. *Ann. Bot. N. Ser.* **31**, 479–487.

Cohen, M. H., and Robertson, A. (1971a). Wave propagation in the early stages of aggregation of cellular slime molds. *J. Theor. Biol.* **31**, 101–118.

Cohen, M. H., and Robertson, A. (1971b). Chemotaxis and the early stages of aggregation in the cellular slime molds. *J. Theor. Biol.* **31**, 119–135.

Cohen, M. H., and Robertson, A. (1971c). Differentiation for aggregation in the cellular slime molds. *Proc. 1st. Int. Conf. Cell Diff.* (R. Harris, D. Viza, eds.). pp. 35–45. Munksgaard, Copenhagen.

Cook, W. R. I. (1939). Some observations on *Sappinia pedata* Dang. *Trans. Brit. Mycol. Soc.* **22**, 302–306.

Coston, M. B., and Loomis, W. F., Jr. (1969). Isozymes of β-glucosidase in *Dictyostelium discoideum. J. Bacteriol.* **100**, 1208–1217.

Cotter, D. A., and Raper, K. B. (1966). Spore germination in *Dictyostelium discoideum. Proc. Nat. Acad. Sci. U.S.* **56**, 880–887.

Cotter, D. A., and Raper, K. B. (1968a). Factors affecting the rate of heat-induced spore germination in *Dictyostelium discoideum. J. Bacteriol.* **96**, 86–92.

Cotter, D. A., and Raper, K. B. (1968b). Properties of germinating spores of *Dictyostelium discoideum. J. Bacteriol.* **96**, 1680–1689.

Cotter, D. A., and Raper, K. B. (1968c). Spore germination in strains of *Dictyostelium discoideum* and other members of the Dictyosteliaceae. *J. Bacteriol.* **96**, 1690–1695.

Cotter, D. A., and Hohl, H. R. (1969). Correlation between plaque size and spore size in *Dictyostelium discoideum. J. Bacteriol.* **98**, 321–322.

Cotter, D. A., and Raper, K. B. (1970). Spore germination in *Dictyostelium discoideum:* Trehalase and the requirement for protein synthesis. *Develop. Biol.* **22**, 112–128.

Cotter, D. A., Miura–Santo, L. Y., and Hohl, H. R. (1969). Ultrastructural changes during germination of *Dictyostelium discoideum* spores. *J. Bacteriol.* **100**, 1020–1026.

Coukell, M. B., and Walker, I. O. (1973). The basic nuclear proteins of the cellular slime mold *Dictyostelium discoideum. Cell. Diff.* **2**, 77–85.

Crick, F. (1970). Diffusion in embryogenesis. *Nature (London)* **225**, 420–423.

Dangeard, P. A. (1896). Contribution à l'étude des Acrasiées. *Botaniste* **5**, 1–20.

Davidoff, F., and Korn, E. D. (1962). Lipids of *Dictyostelium discoideum:* Phospholipid composition and the presence of two new fatty acids, cis,cis-5, 11-octadecadienoic and cis,cis-5,9-hexadecadienoic acids. *Biohem. Biophys. Res. Commun.* **9**, 54–58.

Davidoff, F., and Korn, E. D. (1963a). Fatty acid and phospholipid composition of the cullular slime mold, *Dictyostelium discoideum:* The occurrence of previously undescribed fatty acids. *J. Biol. Chem.* **238**, 3199–3209.

Davidoff, F., and Korn, E. D. (1963b). The biosynthesis of fatty acids in the cellular slime mold, *Dictyostelium discoideum. J. Biol. Chem.* **238**, 3210–3215.

Deering, R. A., Adolf, A. C., and Silver, S. M. (1972). Independence of propagation ability and developmental processes in irradiated cellular slime molds. *Int. J. Rad. Biol.* **21**, 235–245.

Dehaan, R. L. (1959). The effects of the chelating agent ethylenediamine tetra-acetic acid on cell adhesion in the slime mold *Dictyostelium discoideum. J. Embryol. Exp. Morphol.* **7**, 335–343.

Demerec, M., Adelberg, E., Clark, A., and Hartman, P. (1966). A proposal for a uniform nomenclature in bacterial genetics. *Genetics* **54**, 61–67.

Dimond, R., and Loomis, W. F., Jr. (1973). Acetylglucosaminidase mutants in *Dictyostelium discoideum. Genetics* **2**, 64.

Dimond, R., and Loomis, W. F., Jr. (1974). Vegetative isozyme of N-acetylglucosaminidase in *Dictyostelium discoideum. J. Biol. Chem.* **249**, 5628–5632.

Dimond, R., Brenner, M., and Loomis, W. F., Jr. (1973). Mutations affecting *N*-acetylglucosaminidase in *Dictyostelium discoideum. Proc. Nat. Acad. Sci. U.S.* **70**, 3356–3360.

Durston, A. (1973). *Dictyostelium discoideum* aggregation fields as excitable media. *J. Theor. Biol.* **42**, 483–504.

Durston, A. (1974). Pacemaker activity during aggregation in *Dictyostelium discoideum. Develop. Biol.* **37**, 225–235.

Dutta, S. K., and Garber, E. D. (1961). The identification of physiological races of a fungal phytopathogen using strains of the slime mold *Acrasis rosea. Proc. Nat. Acad. Sci. U.S.* **47**, 990–993.

Edmundson, T. D., and Ashworth, J. M. (1972). 6-Phosphogluconate dehydrogenase and the assay of uridine diphosphate glucose pyrophosphorylase in the cellular slime mould *Dictyostelium discoideum. Biochem. J.* **126**, 593–600.

Ellingson, J. S. (1974). Changes in the phospholipid composition in the differentiating cellular slime mold *Dictyostelium discoideum. Biochim. Biophys. Acta* **337**, 60–67.

Ellingson, J. S., Telser, A., and Sussman, M. (1971). Regulation of functionally related enzymes during alternative developmental programs. *Biochim. Biophys. Acta* **244**, 388–395.

Ennis, H. L., and Sussman, M. (1958a). Synergistic morphogenesis by mixtures of *Dictyostelium discoideum* wild-type and aggregateless mutants. *J. Gen. Microbiol.* **18**, 433–449.

Ennis, H. L., and Sussman, M. (1958b). The initiator cell for slime mold aggregation. *Proc. Nat. Acad. Sci. U.S.* **44**, 401–411.

Ennis, H. L., and Sussman, M. (1958c). The initiator cell for slime mold aggregation (Abstract). *Bacteriol. Proc.* 32.

Epstein, R., Bollé, A., Steinberg, C., Kellenberger, E., Boy de la Tour, E., Chevally, R., Edgar, R., Sussman, M., Denhardt, G. H., and Leilausis, A. (1963). Physiological studies of conditional lethal mutations of bacteriophage T4 D. *Cold Spring Harbor Symp. Quant. Biol.* **28**, 375–394.

Erdos, G. W., Nickerson, A. W., and Raper, K. B. (1972). Fine structure of macrocysts in *Polysphondylium violaceum. Cytobiol.* **6**, 351–366.

Erdos, G. W., Nickerson, A. W., and Raper, K. B. (1973a). The fine structure of macrocyst germination in *Dictyostelium mucoroides. Develop. Biol.* **32**, 321–330.

Erdos, G. W., Raper, K. B., and Vogen, L. K. (1973b). Mating types and macrocyst formation in *Dictyostelium discoideum. Proc. Nat. Acad. Sci. U.S.* **70**, 1828–1830.

Erickson, S. K., and Ashworth, J. M. (1969). The mitochondrial electron-transport system of the cellular slime mould *Dictyostelium discoideum. Biochem. J.* **113**, 567–568.

Every, D., and Ashworth, J. M. (1973). The purification and properties of extracellular glycosidases of the cellular slime mould *Dictyostelium discoideum. Biochem. J.* **133**, 37–47.

Farnsworth, P. (1973a). Aspects of growth morphogenesis, and patterns of differentiation in the cellular slime mould *Dictyostelium discoideum.* Thesis, Univ. London.

Farnsworth, P. (1973b). Morphogenesis in the cellular slime mould *Dictyostelium discoideum;* the formation and regulation of aggregate tips and the specification of developmental axes. *J. Embryol. Exp. Morphol.* **29**, 253–266.

Farnsworth, P. (1974). Experimentally induced aberrations in the pattern of differentiation in the cellular slime mould *Dictyostelium discoideum. J. Embryol. Exp. Morphol.* **31**, 435–451.

Farnsworth, P., and James, R. (1972). An effect of the *lon* phenotype in *Escherichia coli* as indicated by the growth of amoebae of *Dictyostelium discoideum. J. Gen. Microbiol.* **73**, 447–454.

Farnsworth, P., and Loomis, W. F., Jr. (1974). A barrier to diffusion in pseudoplasmodia of *Dictyostelium discoideum. Devel. Biol.* **41**, 77–83.

Farnsworth, P., and Wolpert, L. (1971). Absence of cell sorting out in the grex of the slime mould *Dictyostelium discoideum. Nature (London)* **231**, 329–330.

Farrell, C. A., and DeToma, F. J. (1973). Increased capacity for RNA-synthesis in *Dictyostelium discoideum* nuclei by exposure to cyclic AMP or 5′-AMP. *Biochem. Biophys. Res. Commun.* **54**, 1504–1510.

Faust, R. G., and Filosa, M. F. (1959). Permeability studies on the amoebae of the slime mold, *Dictyostelium mucoroides. J. Cell. Comp. Physiol.* **54**, 297–298.

Fawcett, D. W. (1966). "An Atlas of Fine Structure" Saunders, Philadelphia, Pennsylvania.

Fayod, V. (1883). Beitrag zur kenntnis niederer myxomyceten. *Botan. Zeitung* **41**, 169–177.

Ferber, E., Munder, P. G., Fischer, H., and Gerisch, G. (1970). High phospholipase activities in amoebae of *Dictyostelium discoideum. Eur. J. Biochem.* **14**, 253–257.

Filosa, M. F. (1960). The effects of ethionine on the morphogenesis of cellular slime molds (abstract). *Anat. Rec.* **138**, 348.

Filosa, M. F. (1962). Heterocytosis in cellular slime molds. *Amer. Natur.* **96**, 79–91.

Filosa, M. F., and Dengler, R. E. (1972). Ultrastructure of macrocyst formation in the cellular slime mold, *Dictyostelium mucoroides:* Extensive phagocytosis of amoebae by a specialized cell. *Develop. Biol.* **29**, 1–16.

Firtel, R. A. (1972). Changes in the expression of single-copy DNA during development of the cellular slime mold *Dictyostelium discoideum. J. Mol. Biol.* **66**, 363–377.

Firtel, R. A., and Bonner, J. T. (1970). Developmental control of alpha 1-4 glucan phosphorylase in cellular slime mold *Dictyostelium discoideum. Proc. Fed. Amer. Soc. Exp. Biol.* **29**, 669.

Firtel, R. A., and Bonner, J. T. (1972a). Characterization of the genome of the cellular slime mold *Dictyostelium discoideum. J. Mol. Biol.* **66**, 339–361.

Firtel, R. A., and Bonner, J. T. (1972b). Developmental control of alpha 1-4 glucan phosphorylase in the cellular slime mold *Dictyostelium discoideum. Develop. Biol.* **29**, 85–103.

Firtel, R. A., and Brackenbury, R. W. (1972). Partial characterization of several protein and amino acid metabolizing enzymes in the cellular slime mold *Dictyostelium discoideum. Develop. Biol.* **27**, 307–321.

Firtel, R. A., and Lodish, H. F. (1973). A small nuclear precursor of messenger RNA in the cellular slime mold *Dictyostelium discoideum. J. Mol. Biol.* **79**, 295–314.

Firtel, R. A., Jacobson, A., and Lodish, H. F. (1972). Isolation and hybridization kinetics of messenger RNA from *Dictyostelium discoideum. Nature (London) New Biol.* **239**, 225–228.

Firtel, R. A., Baxter, L., and Lodish, H. F. (1973). Actinomycin D and the regulation of enzyme biosynthesis during development of *Dictyostelium discoideum. J. Mol. Biol.* **79**, 315–327.

Firtel, R. A., Jacobson, A., Tuchman, J., and Lodish, H. F. (1974). Gene activity during development of the cellular slime mold *Dictyostelium discoideum. Genetics* (in press).

Firtel, R. A., Kindle, K., and Huxley, M. P. (1975). Structural organization and processing of the genetic transcript in the cellular slime mold *Dictyostelium discoideum*. *Fed. Proc. Fed. Amer. Soc. Biol.* (in press).

Firtel, R. A., and Lodish, H. F., in press.

Francis, D. W. (1959). Pseudoplasmodial movement in *Dictyostelium discoideum*. M.S. Thesis, Univ. Wisconsin, Madison, Wisconsin.

Francis, D. W. (1962). The movement of pseudoplasmodia of *Dictyostelium discoideum*. Ph.D. Thesis, Univ. Wisconsin, Madison, Wisconsin.

Francis, D. W. (1964). Some studies on phototaxis of *Dictyostelium*. *J. Cell. Comp. Physiol.* **64**, 131–138.

Francis, D. W. (1965). Acrasin and the development of *Polysphondylium pallidum*. *Develop. Biol.* **12**, 329–346.

Francis, D. W. (1969). Time sequences for differentiation in cellular slime molds. *Quart. Rev. Biol.* **44**, 277–290.

Francis, D. W., and O'Day, D. H. (1971). Sorting out in pseudoplasmodia of *Dictyostelium discoideum*. *J. Exp. Zool.* **176**, 265–272.

Franke, J., and Sussman, M. (1971). Synthesis of uridine diphosphate glucose pyrophosphorylase during the development of *Dictyostelium discoideum*. *J. Biol. Chem.* **246**, 6381–6388.

Franke, J., and Sussman, M. (1973). Accumulation of uridine diphosphoglucose pyrophosphatase in *Dictyostelium discoideum* via preferential synthesis. *J. Mol. Biol.* **81**, 173–185.

Free, S. J., and Loomis, W. F., Jr. (1975). Isolation of mutations in *Dictyostelium discoideum* affecting α-mannosidase. *Biochimie*, in press.

Freeze, H., and Loomis, W. F., Jr., in preparation.

Freim, J. O., Jr., and Deering, R. A. (1970). Ultraviolet irradiation of the vegetative cells of *Dictyostelium discoideum*. *J. Bacteriol.* **102**, 36–42.

Fukui, Y., and Takeuchi, I. (1971). Drug resistant mutants and appearance of heterozygotes in the cellular slime mould *Dictyostelium discoideum*. *J. Gen. Microbiol.* **67**, 307–317.

Fuller, M. S., and Rakatansky, R. M. (1966). A preliminary study of the carotenoids in *Acrasis rosea*. *Can. J. Bot.* **44**, 269–274.

Garrod, D. R. (1969a). The cellular basis of movement of the migrating grex of the slime mould *Dictyostelium discoideum*. *J. Cell Sci.* **4**, 781–798.

Garrod, D. R. (1969b). The way some cells creep. *New Sci.* August 285–287.

Garrod, D. R. (1972). Acquisition of cohesiveness by slime mould cells prior to morphogenesis. *Exp. Cell Res.* **72**, 588–591.

Garrod, D. R., and Ashworth, J. M. (1972). Effect of growth conditions on development of the cellular slime mould, *Dictyostelium discoideum*. *J. Embryol. Exp. Morphol.* **28**, No. 2, 463–479.

Garrod, D. R., and Ashworth, J. M. (1973). Development of the cellular slime mould *Dictyostelium discoideum*. *Symp. Soc. Gen. Microbiol.* **23**, 407–435.

Garrod, D. R., and Born, G. V. R. (1971). Effect of temperature on the mutual adhesion of preaggregation cells of the slime mould, *Dictyostelium discoideum*. *J. Cell Sci.* **8**, 751–765.

Garrod, D. R., and Gingell, D. (1970). A progressive change in the electrophoretic mobility of preaggregation cells of the slime mould, *Dictyostelium discoideum*. *J. Cell Sci.* **6**, 277–284.

Garrod, D. R., and Malkinson, A. (1973). Cyclic AMP, pattern formation and movement in the slime mould, *Dictyostelium discoideum*. *Exp. Cell Res.* **81**, 492–495.

Garrod, D. R., and Wolpert, L. (1968). Behaviour of the cell surface during movement of pre-aggregation cells of the slime mould *Dictyostelium discoideum* studied with fluorescent antibody. *J. Cell Sci.* 3, 365–372.

Garrod, D. R., Palmer, J. F., and Wolpert, L. (1970). Electrical properties of the slime mould grex. *J. Embryol. Exp. Morphol.* 23, 311–322.

George, R. (1968). Cell organization and ultrastructure during culmination of cellular slime molds. Ph.D. Thesis, Univ. of Hawaii.

George, R. P., Albrecht, R. M., Raper, K. B., Sachs, I. B., and MacKenzie, A. P. (1970). Rapid freeze preparation of *Dictyostelium discoideum* for scanning electron microscopy. *Proc. Cambridge. Stereoscan Coll.*, 159–165.

George, R. P., Albrecht, Raper, K. B., Sachs, I. B., and MacKenzie, A. P. (1972a). Scanning electron microscopy of spore germination in *Dictyostelium discoideum*. *J. Bacteriol.* 112, 1383–1386.

George, R. P., Hohl, H. R., and Raper, K. (1972b). Ultrastructural development of stalk-producing cells in *Dictyostelium discoideum*, a cellular slime mould. *J. Gen. Microbiol.* 70, 477–489.

Gerisch, G. (1959). Ein Submerskulturverfahren für entwicklungsphysiologische untersuchungen an *Dictyostelium discoideum*. *Naturwissenschaften* 46, 654–656.

Gerisch, G. (1960). Zellfunktionen und Zellfunktionswechsel in der Entwicklung von *Dictyostelium discoideum*. I. Zellagglutination und induktion der fruchtkörperpolarität. *Wilhelm Roux' Arch Entwicklungsmech. Organismen* 152, 632–654.

Gerisch, G. (1961a). Zellfunktionen und Zellfunktionswechsel in der Entwicklung von *Dictyostelium discoideum*. II. Aggregation homogener Zellpopulationen und Zentrenbildung. *Develop. Biol.* 3, 685–724.

Gerisch, G. (1961b). Zellfunktionen und Zellfunktionswechsel in der Entwicklung von *Dictyostelium discoideum*. III. Gentrennte Beeinflussung von Zelldifferenzierung und Morphogenese. *Wilhelm Roux Arch. Entwicklungsmech. Organismen* 153, 158–167.

Gerisch, G. (1961c). Zellfunktionen und Zellfunktionswechsel in der Entwicklung von *Dictyostelium discoideum*. V. Stadienspezifische Zelkontaktbidung und ihre quantitative Erfassung. *Exp. Cell. Res.* 25, 535–554.

Gerisch, G. (1961d). Zellkontaktbildung vegetativer und aggregationsreifer Zellen von *Dictyostelium discoideum*. *Naturwissenschaften* 48, 436–437.

Gerisch, G. (1962a). Die zellulären schleimpilze als objekte der entwicklungsphysiologie. *Ber. D. Bot. Ges.* 75, 82–89.

Gerisch, G. (1962b). Zellfunktionen und Zellfunktionswechsel in der Entwicklung von *Dictyostelium discoideum*. IV. Der Zeitplan der Entwicklung. *Wilhelm Roux Arch. Entwicklungsmech. Organismen* 153, 603–620.

Gerisch, G. (1962c). Zellfunktionen und Zellfunktionswechsel in der Entwicklung von *Dictyostelium discoideum*. VI. Inhibitoren der aggregation, ihr einfluss auf zellkontaktbildung und morphogenetische bewegung. *Exp. Cell Res.* 26, 462–484.

Gerisch, G. (1963). Eine für *Dictyostelium* ungewöhnliche aggregationsweise. *Naturwissenschaften.* 50, 160–161.

Gerisch, G. (1964a). Entwicklung von *Dictyostelium* [Commentary on a film]. *Publikationen zu Wissenschaftlichen Filmen.* I, 127–140.

Gerisch, G. (1964b). Die Bildung des Zellverbandes bei *Dictyostelium minutum*. I. Übersicht über die Aggregation und den Funktionswechsel der Zellen. *Wilhelm Roux Arch. Entwicklungsmech. Organismen* 155, 342–357.

Gerisch, G. (1964c). *Dictyostelium discoideum* (Acrasina). Aggregation und bildung des sporophors. *Pub. Wiss. Filmen* 1, 255–264.

Gerisch, G. (1964d). *Dictyostelium minutum* (Acrasina). Aggregation. *Publ. Wiss. Filmen* 1, 265–278.

Gerisch, G. (1964e). Eine mutante von *Dictyostelium minutum* mit blockierter zentrengrundung. *Z. Naturforsch. B* **20**, 298–301.

Gerisch, G. (1964f). Stadienspezifische aggregationsmuster von *Dictyostelium discoideum*. *Wilhelm Roux Arch. Entwicklungsmech. Organismen* **156**, 127–144.

Gerisch, G. (1964g). Spezifische zellkontakte als mechanismen der tierischen entwicklung. *Umschau* **65**, 392–395.

Gerisch, G. (1966). Die bildung des zellverbandes bei *Dictyostelium minutum*. II. Analyse der zentrengrundung anhand von filmaufnahmen. *Wilhelm Roux Arch. Entwicklungsmech. Organismen* **157**, 174–189.

Gerisch, G. (1968). Cell aggregation and differentiation in *Dictyostelium*. *In* "Current Topics in Developmental Biology" (A. Moscana and A. Monroy, eds.), Vol. 3, pp. 157–197. Academic Press, New York.

Gerisch, G. (1970). Immunchemische untersuchungen an plasmamembranen aggregierender zellen. *Deut. Z. Gesell.* **64**, 6–14.

Gerisch, G. (1971). Periodische signale steuern die musterbildung in zellverbanden. *Naturwissenschaften* **58**, 430–438.

Gerisch, G., and Hess, B. (1973). Cyclic AMP-controlled oscillations in suspended *Dictyostelium* cells: Their relation to morphogenetic cell interactions. *Hoppe-Seyler's Z. Physiol. Chem.* **354**, 1193.

Gerisch, G., and Hess, B. (1974). Cyclic AMP-controlled oscillations in suspended *Dictyostelium* cells: Their relation to morphogenetic cell interactions. *Proc. Nat. Acad. Sci. U.S.* **71**, 2118–2122.

Gerisch, G., Normann, I., and Beug, H. (1966). Rhythmik der zellorientierung und der bewegungsgeschwindigkeit im chemotaktischen readtionssystem von *Dictyostelium discoideum*. *Naturwissenschaften* **23**, 618.

Gerisch, G., Lüderitz, O., and Ruschmann, E. (1967). Antikörper fördern die phagozytose von bakterien durch amöben. *Z. Naturforsch.* **22**, 109.

Gerisch, G., Malchow, D., Wilhelms, H., and Lüderitz, O. (1969). Artspezifität polysaccharid–haltiger zellmembran–antigene von *Dictyostelium discoideum*. *Eur. J. Biochem.* **9**, 229–236.

Gerisch, G., Riedel, V., and Malchow, D. (1971). Zyklisches adenosinmonophosphat als signalstoff der entwicklung. *Umschau* **14**, 532–533.

Gerisch, G., Malchow, D., Riedel, V., Müller, E., and Every, M. (1972). Cyclic AMP phosphodiesterase and its inhibitor in slime mould development. *Nature (London) New Biol.* **235**, 90–92.

Gerisch, G., Beug, H., Malchow, D., Schwarz, H., and Stein, A. (1974). Receptors for intercellular signals in aggregating cells of the slime mold, *Dictyostelium discoideum*. *In* "Biology and Chemistry of Eukaryotic Cell Surfaces."

Gezelius, K. (1959). The ultrastructure of cells and cellulose membranes in Acrasiae. *Exp. Cell Res.* **18**, 425–453.

Gezelius, K. (1961). Further studies in the ultrastructure of Acrasiae. *Exp. Cell Res.* **23**, 300–310.

Gezelius, K. (1962). Growth of the cellular slime mold *Dictyostelium discoideum* on dead bacteria in liquid media. *Physiol. Plant.* **15**, 587–592.

Gezelius, K. (1971). Acid phosphatase localization in myxamoebae of *Dictyostelium discoideum*. *Arch. Mikrobiol.* **75**, 327–337.

Gezelius, K. (1972). Acid phosphatase localization during differentiation in the cellular slime mold *Dictyostelium discoideum*. *Arch. Mikrobiol.* **85**, 51–76.

Gezelius, K., and Rånby (1957). Morphology and fine structure of the slime mold *Dictyostelium discoideum. Exp. Cell. Res.* 12, 265–289.

Gezelius, K., and Wright, B. E. (1965). Alkaline phosphatase in *Dictyostelium discoideum. J. Gen. Microbiol.* 38, 309–327.

Gillette, M. U., and Filosa, M. F. (1973). Effect of concanavalin A on cellular slime mold development: premature appearance of membrane-bound cyclic AMP phosphodiesterase. *Biochem. Biophys. Res. Commun.* 53, 1159–1166.

Gingell, D. (1971). Computed force and energy of membrane interaction. *J. Theor. Biol.* 30, 121–149.

Gingell, D., and Garrod, D. R. (1969). Effect of EDTA on electrophoretic mobility of slime mould cells and its relationship to current theories of cell adhesion. *Nature (London)* 221, 192–193.

Gingell, D., Garrod, D. R., and Palmer, J. F. (1969). Divalent cations and cell adhesion. "Symposium on Calcium and Cell Function" (A. Cuthbert, ed.), pp. 59–64. Macmillan, New York.

Gingold, E., and Ashworth, J. M. (1974). Evidence for mitotic crossing-over during the parasexual cycle of the cellular slime mold *Dictyostelium discoideum., J. Gen. Microbiol.* 84, 70–78.

Goidl, E. A., Chassy, B. M., Love, L. L., and Krichevsky, M. I. (1972). Inhibition of aggregation and differentiation of *Dictyostelium discoideum* by antibodies against adenosine 3′:5′-cyclic monophosphate diesterase. *Proc. Nat. Acad. Sci. U.S.* 69, 1128–1130.

Goldstone, E. M., Banerjee, S., Allen, J. R., Lee, J. J., Hutner, S. H., Bacchi, C. J., and Melville, J. F. (1966). Minimal defined media for vegetative growth of the acrasian *Polysphondylium pallidum* WS-320. *J. Protozool.* 13, 171–174.

Green, A., and Newell, P. (1974). The isolation and subfractionation of plasma membrane from the cellular slime mould *Dictyostelium discoideum. Biochem. J.* 140, 313–322.

Gregg, J. H. (1950). Oxygen utilization in relation to growth and morphogenesis of the slime mold *Dictyostelium discoideum. J. Exp. Zool.* 114, 173–196.

Gregg, J. H. (1956). Serological investigations of cell adhesion in the slime molds, *Dictyostelium discoideum, D. purpureum,* and *Polysphondylium violaceum. J. Gen. Physiol.* 39, 813–820.

Gregg, J. H. (1957). Serological investigations of aggregateless variants of the slime mold, *Dictyostelium discoideum* (abstract). *Anat. Rec.* 128, 558.

Gregg, J. H. (1960). Surface antigen dynamics in the slime mold, *Dictyostelium discoideum. Biol. Bull.* 118, 70–78.

Gregg, J. H. (1961). An immunoelectrophoretic study of the slime mold *Dictyostelium discoideum. Develop. Biol.* 3, 757–766.

Gregg, J. H. (1964). Developmental processes in cellular slime molds. *Physiol. Rev.* 44, 631–656.

Gregg, J. H. (1965a). Centrifugal homogenizer. *Science* 150, 1739–1740.

Gregg, J. H. (1965b). Regulation in the cellular slime molds. *Develop. Biol.* 12, 377–393.

Gregg, J. H. (1966a). Organization and synthesis in the cellular slime molds. *In* "The Fungi, An Advanced Treatise" (G. C. Ainsworth, and A. S. Sussman, eds.), Vol. 2, pp. 235–281, Academic Press, New York.

Gregg, J. H. (1966b). A microrespirometer capable of quantitative substrate mixing. *Exp. Cell Res.* 42, 260–264.

Gregg, J. H. (1967). Cellular slime molds. *In* "Techniques for the Study of Development" (F. Wilt and N. Wessells, eds.), pp. 359–376, Crowell–Collier, New York.

Gregg, J. H. (1968). Prestalk cell isolates in *Dictyostelium. Exp. Cell Res.* **51**, 633–642.

Gregg, J. H. (1971). Developmental potential of isolated *Dictyostelium* myxamoebae. *Develop. Biol.* **26**, 478–485.

Gregg, J. H., and Aldrich, H. C. (1972). Unit membrane structural changes following cell association in *Dictyostelium. J. Cell. Biol.* **55**, 95a.

Gregg, J. H., and Badman, W. S. (1970). Morphogenesis and ultrastructure in *Dictyostelium. Develop. Biol.* **22**, 96–111.

Gregg, J. H., and Badman, W. S. (1973). Transitions in differentiation by the cellular slime molds. *In* "Developmental Regulation" (J. Stuart, ed.), Academic Press, New York.

Gregg, J. H., and Bronsweig, R. D. (1954). The carbohydrate metabolism of the slime mold, *Dictyostelium discoideum,* during development (abstract). *Biol. Bull.* **107**, 312.

Gregg, J. H., and Bronsweig, R. D. (1956a). Dry weight loss during culmination of the slime mold, *Dictyostelium discoideum. J. Cell. Comp. Physiol.* **47**, 483–488.

Gregg, J. H., and Bronsweig, R. D. (1956b). Biochemical events accompanying stalk formation in the slime mold, *Dictyostelium discoideum. J. Cell. Comp. Physiol.* **48**, 293–300.

Gregg, J. H., and Nesom, M. G. (1973). Response of *Dictyostelium* plasma membranes to adenosine 3′:5′-cyclic monophosphate. *Proc. Nat. Acad. Sci. U.S.* **70**, 1630–1633.

Gregg, J. H., and Trygstad, C. W. (1958). Surface antigen defects contributing to developmental failure in aggregateless variants of the slime mold, *Dictyostelium discoideum. Exp. Cell Res.* **15**, 358–369.

Gregg, J. H., Hackney, A. L., and Krivanek, J. O. (1954). Nitrogen metabolism of the slime mold *Dictyostelium discoideum* during growth and morphogenesis. *Biol. Bull.* **107**, 226–235.

Gustafson, G. L., and Wright, B. E. (1971). UDP-glucose pyrophosphorylase from the cellular slime mold (Abst. #98) *Fed. Proc. Fed. Amer. Soc. Exp. Biol.* **30**, 1069.

Gustafson, G. L., and Wright, B. E. (1972). Analysis of approaches used in studying differentiation of the cellular slime mold. *Crit. Rev. Microbiol.,* pp. 453–478.

Gustafson, G. L., and Wright, B. E. (1973a). UDP-glucose pyrophosphorylase synthesis in myxamoebae of *Dictyostelium discoideum. Biochem. Biophys. Res. Commun.* **50**, 438–442.

Gustafson, G. L., and Wright, B. E. (1973b). Accumulation of UDP glucose pyrophosphorylase during differentiation of *Dictyostelium discoideum.* (abstract) *Fed. Proc. Fed. Amer. Soc. Exp. Biol.* **32**, 652.

Gustafson, G. L., Kong, W. Y., and Wright, B. E. (1973). Analysis of uridine diphosphate–glucose pyrophosphorylase synthesis during differentiation in *Dictyostelium discoideum. J. Biol. Chem.* **248**, 5188–5196.

Hames, B. D., and Ashworth, J. M. (1974a). The metabolism of macromolecules during the differentiation of myxamoebae of the cellular slime mould *Dictyostelium discoideum* containing different amounts of glycogen. *Biochem. J.* **142**, 301–316.

Hames, B. D., and Ashworth, J. M. (1974b). The control of saccharide synthesis during development of myxamoebae of *Dictyostelium discoideum* containing differing amounts of glycogen. *Biochem. J.* **142**, 317–329.

Hames, B. D., Weeks, G., and Ashworth, J. M. (1972). Glycogen synthetase and the control of glycogen synthesis in the cellular slime mould *Dictyostelium discoideum* during cell differentiation. *Biochem. J.* 126, 627–633.

Hammond, J. (1973). A membrane component of the cellular slime mold *Dictyostelium discoideum* rapidly labelled with phosphorus-32 orthophosphate. *Biochim. Biophys. Acta* 291, 371–387.

Harper, R. A. (1926). Morphogenesis in *Dictyostelium. Bull. Torrey Botan. Club* 53, 229–268.

Harper, R. A. (1929). Morphogenesis in *Polysphondylium. Bull. Torrey Botan. Club* 56, 227–258.

Harper, R. A. (1932). Organization and light relations in *Polysphondylium. Bull. Torrey Botan. Club* 59, 49–84.

Harrington, B. J., and Raper, K. B. (1968). Use of a fluorescent brightener to demonstrate cellulose in the cellular slime molds. *J. Appl. Microbiol.* 16, 106–113.

Hashimoto, Y. (1971). Effect of radiation on the germination of spores of *Dictyostelium discoideum. Nature* (*London*) 231, 316–317.

Heftmann, E., Wright, B. E., and Liddel, G. U. (1959). Identification of a sterol with acrasin activity in a slime mold. *J. Amer. Chem. Soc.* 81, 6525.

Heftmann, E., Wright, B. E., and Liddel, G. U. (1960). The isolation of delta^{22}-stigmasten-3 beta-ol from *Dictyostelium discoideum. Arch. Biochem. Biophys.* 91, 266–270.

Hemmes, D. E., Kojima–Buddenhagen, E. S., and Hohl, H. R. (1972). Structural and enzymic analysis of the spore wall layers in *Dictyostelium discoideum. J. Ultrastr. Res.* 41, 406–417.

Hirschberg, E. (1955). Some contributions of microbiology to cancer research. *Bacteriol. Rev.* 19, 65–78.

Hirschberg, E., and Merson, G. (1955). Effect of test compounds on the aggregation and culmination of the slime mold *Dictyostelium discoideum. Cancer Res. Suppl.* 3, 76–79.

Hirschberg, E., and Rusch, H. P. (1950). Effects of compounds of varied biochemical action on the aggregation of a slime mold, *Dictyostelium discoideum. J. Cell. Comp. Physiol.* 36, 105–113.

Hirschberg, E., and Rusch, H. P. (1951). Effect of 2,4-dinitrophenol on the differentiation of the slime mold *Dictyostelium discoideum. J. Cell. Comp. Physiol.* 37, 323–336.

Hirschberg, E., Ceccarini, C., Osnos, M., and Carchman, R. (1968). Effects of inhibitors of nucleic acid and protein synthesis on growth and aggregation of the cellular slime mold *Dictyostelium discoideum. Proc. Nat. Acad. Sci. U.S.* 61, 316–323.

Hirschy, R. A., and Raper, K. B. (1964). Light control of macrocyst formation in *Dictyostelium* (abstract). *Bacteriol. Proc.* 27.

Hohl, H. R. (1965). Nature and development of membrane systems in food vacuoles of cellular slime molds predatory upon bacteria. *J. Bacteriol.* 90, 755–765.

Hohl, H. R., and Hamamoto, S. T. (1967). Reversal of ethionine inhibition by methionine during slime mold development. *Pacific Sci.* 21, 534–538.

Hohl, H. R., and Hamamoto, S. T. (1969a). Ultrastructure of *Acrasis rosea,* cellular slime mold during development. *J. Protozool.* 16, 333–344.

Hohl, H. R., and Hamamoto, S. T. (1969b). Ultrastructure of spore differentiation in *Dictyostelium:* The prespore vacuole. *J. Ultrastr. Res.* 26, 442–453.

Hohl, H. R., and Jehli, J. (1973). The presence of cellulose microfibrils in the proteinaceous slime track of *Dictyostelium discoideum. Arch. Mikrobiol.* 92, 179–187

Hohl, H. R., and Raper, K. B. (1963a). Nutrition of cellular slime molds. I. Growth on living and dead bacteria. *J. Bacteriol.* **85**, 191–198.

Hohl, H. R., and Raper, K. B. (1963b). Nutrition of cellular slime molds. II. Growth of *Polysphondylium pallidum* in axenic culture. *J. Bacteriol.* **85**, 199–206.

Hohl, H. R., and Raper, K. B. (1963c). Nutrition of cellular slime molds. III. Specific growth requirements of *Polysphondylium pallidum*. *J. Bacteriol.* **86**, 1314–1320.

Hohl, H. R., and Raper, K. B. (1964). Control of sorocarp size in the cellular slime mold *Dictyostelium discoideum*. *Develop. Biol.* **9**, 137–153.

Hohl, H. R., Hamamoto, S. T., and Hemmes, D. E. (1968). Ultrastructural aspects of cell elongation, cellulose synthesis, and spore differentiation in *Actyostelium leptosomum*, a cellular slime mold. *Amer. J. Bot.* **55**, 783–796.

Huffman, D. M., and Olive, L. S. (1963). A significant morphogenetic variant of *Dictyostelium mucoroides*. *Mycologia* **55**, 333–344.

Huffman, D. M., and Olive, L. S. (1964). Engulfment and anastomosis in the cellular slime molds (Acrasiales). *Amer. J. Bot.* **51**, 465–471.

Huffman, D. M., Kahn, A. J., and Olive, L. S. (1962). Anastomosis and cell fusions in *Dictyostelium*. *Proc. Nat. Acad. Sci. U.S.* **48**, 1160–1164.

Huxley, H. (1973). Muscular contraction and cell mobility. *Nature* (*London*) **243**, 445–449.

Ikeda, T., and Takeuchi, I. (1971). Isolation and characterization of a prespore specific structure of the cellular slime mold, *Dictyostelium discoideum*. *Develop. Growth and Differentiation* **13**, 221–229.

Ishida, S., Maeda, Y., and Takeuchi, I. (1974). An anucleolate mutant of the cellular slime mold *Dictyostelium discoideum*. *J. Gen. Microbiol.* **81**, 491–499.

Ito, K., and Iwabuchi, M. (1971). Conformational changes of ribosomal subunits of *Dictyostelium discoideum* by salts. *Biochem. J.* **69**, 1135–1138.

Iwabuchi, M., and Ochiai, H. (1969). Sedimentation properties of ribosomal particles in *Dictyostelium discoideum*. *Biochim. Biophys. Acta* **190**, 211–213.

Iwabuchi, M., Mizukami, Y., and Sameshima, S. (1970a). Synthesis of precursor molecules of ribosomal RNA in the cellular slime mold *Dictyostelium discoideum*. *Biochim. Biophys. Acta* **228**, 693–700.

Iwabuchi, M., Ito, K., and Ochiai, H. (1970b). Characterization of ribosomes in cellular slime mold, *Dictyostelium discoideum*. *J. Biochem. Tokyo,* **68**, 549–556.

Jacob, F., and Monod, J. (1961). On the regulation of gene activity. *Cold Spring Harbor Symp. Quant. Biol.* **26**, 193–212.

Jacobson, A., Firtel, R. A., and Lodish, H. F. (1974a). Synthesis of messenger and ribosomal RNA precursors in isolated nuclei of the cellular slime mold, *Dictyostelium discoideum*. *J. Mol. Biol.* **82**, 213–230.

Jacobson, A., Firtel, R. A., and Lodish, H. F. (1974b). Transcription of poly(dT) sequences in the genome of the cellular slime mold, *Dictyostelium discoideum*. *Proc. Nat. Acad. Sci. U.S.* **71**, 1607–1611.

Jeon, K. W., and Bell, L. G. E. (1964). Behaviour of cell membrane in relation to locomotion in *Amoeba proteus*. *Exp. Cell Res.* **33**, 531–539.

Johnson, D. F., Wright, B. E., and Heftmann, E. (1962). Biogenesis of delta^{22}-stigmasten-3 beta-ol in *Dictyostelium discoideum*. *Arch. Biochem. Biophys.* **97**, 232–235.

Jones, P. C. T. (1969). Temperature and anesthetic induced alterations of ATP level in animal and plant cells, and their biological significance. *Cytobios,* **1B**, 65–71.

Jones, P. C. T. (1970a). The effect of light, temperature, and anaesthetics on ATP levels in the leaves of *Chenopodium rubrum* and *Phaseolus vulgaris*. *J. Exp. Bot.* **21**, 58–63.

Jones, P. C. T. (1970b). The interaction of light and temperature in determining ATP levels in the myxamoebae of the cellular slime mould *Dictyostelium discoideum* Acr. 12. *Cytobios,* **6,** 89–94.

Jones, P. C. T. (1972). Central role for ATP in determining some aspects of animal and plant cell behaviour. *J. Theor. Biol.* **34,** 1–13.

Jones, T. H. D., and Wright, B. E. (1970). Partial purification and characterization of glycogen phosphorylase from *Dictyostelium discoideum. J. Bacteriol.* **104,** 754–761.

Kahn, A. J. (1964a). Some aspects of cell interaction in the development of the slime mold *Dictyostelium purpureum. Develop. Biol.* **9,** 1–19.

Kahn, A. J. (1964b). The influence of light on cell aggregation in *Polysphondylium pallidum. Biol. Bull.* **127,** 85–96.

Kananishi, N., and Watanabe, M. (1973). Radiation resistance in the cellular slime mold. III. Assay of caffeine sensitive recovery of gamma irradiation damage. *J. Radiat. Res.* **14,** 83.

Kanda, F., Ochiai, H., and Iwabuchi, M. (1974). Molecular weight determinations and stoichiometric measurements of 40S and 60S ribosomal proteins of the cellular slime mold *Dictyostelium discoideum. Eur. J. Biochem.* **44,** 469–480.

Katz, E. R., and Bourguignon, L. (1974). The cell cycle and its relationship to aggregation in the cellular slime mold, *Dictyostelium discoideum. Develop. Biol.* **36,** 82–87.

Katz, E. R., and Sussman, M. (1972). Parasexual recombination in *Dictyostelium discoideum:* selection of stable diploid heterozygotes and stable haploid segregants. *Proc. Nat. Acad. Sci. U.S.* **69,** 495–498.

Keller, E. F., and Segal, L. A. (1970a). Initiation of slime mold aggregation viewed as an instability. *J. Theor. Biol.* **26,** 399–415.

Keller, E. F., and Segal, L. A. (1970b). Conflict between positive and negative feedback as an explanation for the initiation of aggregation in slime mould amoebae. *Nature (London)* **277,** 1365–1366.

Kessin, R. (1973). RNA metabolism during vegative growth and morphogenesis of the cellular slime mold, *Dictyostelium discoideum. Develop. Biol.* **31,** 242–251.

Kessin, R., and Newell, P. (1974). Isolation of germination mutants of *Dictyostelium discoideum. J. Bacteriol.* **117,** 379–381.

Kessin, R., Williams, K., and Newell, P. (1974). Linkage analysis in *Dictyostelium discoideum* using temperature sensitive growth mutants selected with bromodeoxyuridine. *J. Bacteriol.* **119,** 776–783.

Kessler, D., and Raper, K. B. (1960). *Guttlina,* a rediscovered genus of cellular slime mold (abstract). *Bacteriol. Proc.* 58.

Khoury, A. T., Deering, R. A., Levin, G., and Altman, G. (1970). Gamma-ray-induced spore germination of *Dictyostelium discoideum. J. Bacteriol.* **104,** 1022–1023.

Killick, K., and Wright, B. E. (1972a). Trehalose synthesis during differentiation in *Dictyostelium discoideum.* III. *In vitro* unmasking of trehalose 6-phosphate synthetase. *J. Biol. Chem.* **247,** 2967–2969.

Killick, K. A., and Wright, B. E. (1972b). Trehalose synthesis during differentiation in *Dictyostelium discoideum.* IV. Secretion of trehalase and the *in vitro* expression of trehalose-6-phosphate synthetase activity. *Biochem. Biophys. Res. Commun.* **48,** 1476–1481.

Killick, K. A., and Wright, B. E. (1974). Regulation of enzyme activity during differentiation in *Dictyostelium discoideum. Annu. Rev. Microbiol.* **28,** 139–166.

Kitzke, E. D. (1952). A new method for isolating members of the Acrasieae from soil samples. *Nature* (*London*) 170, 284–285.

Klein, C. (1974). Presence of magic spot in *Dictyostelium discoideum*. *FEBS Lett.* 38, 149–152.

Konijn, T. M. (1966a). Chemotaxis in the cellular slime molds. I. The effect of temperature. *Develop. Biol.* 12, 487–497.

Konijn, T. M. (1966b). Chemotaxis in the cellular slime molds. II. The effect of density. *Biol. Bull.* 134, 298–304.

Konijn, T. M. (1969). Effect of bacteria on chemotaxis in the cellular slime molds. *J. Bacteriol.* 99, 503–509.

Konijn, T. M. (1970). Microbiological assay of cyclic 3′,5′-AMP. *Experimentia* 26, 367–369.

Konijn, T. M. (1972a). Cyclic AMP as a first messenger. *Advan. Cyclic Nucleotide Res.* 1, 17–31.

Konijn, T. M. (1972b). Cyclic AMP and cell aggregation in the cellular slime molds. *Acta Protozoologica* 11, 137–143.

Konijn, T. M. (1973). The chemotactic effect of cyclic nucleotides with substitutions in the base ring. *FEBS Lett.* 34, 263–266.

Konijn, T. M., and Jastorff, B. (1973). The chemotactic effect of 5′-amido analogues of adenosine cyclic 3′,5′-monosphosphate in the cellular slime moulds. *Biochim. Biophys. Acta* 304, 774–780.

Konijn, T. M., and Raper, K. B. (1961). Cell aggregation in *Dictyostelium discoideum*. *Develop. Biol.* 3, 725–756.

Konijn, T. M., and Raper, K. B. (1965). The influence of light on the time of cell aggregation in the Dictyosteliaceae. *Biol. Bull.* 128, 392–400.

Konijn, T. M., Van de Meene, J. G. C., Bonner, J. T., and Barkley, D. S. (1967). The acrasin activity of adenosine-3′,5′-cyclic phosphate. *Proc. Nat. Acad. Sci. U.S.* 58, 1152–1154.

Konijn, T. M., Barkley, D. S., Chang, Y. Y., and Bonner, J. T. (1968). Cyclic AMP: A naturally occurring acrasin in the cellular slime molds. *Amer. Natur.* 102, 225–233.

Konijn, T. M., Chang, Y. Y., and Bonner, J. T. (1969a). Synthesis of cyclic AMP in *Dictyostelium discoideum* and *Polysphondylium pallidum*. *Nature* (*London*) 224, 1211–1212.

Konijn, T. M., Van de Meene, J. G. C., Change, Y. Y., Barkley, D. S., and Bonner, J. T. (1969b). Identification of adenosine-3′,5′-monophosphate as the bacterial attractant for myxamoebae of *Dictyostelium discoideum*. *J. Bacteriol.* 99, 510–512.

Korn, E., and Wright, P. (1973). Macromolecular composition of an amoeba plasma membrane. *J. Biol. Chem.* 248, 439.

Krichevsky, M. I., and Love, L. L. (1964a). The uptake and utilization of histidine by washed amoebae in the course of development in *Dictyostelium discoideum*. *J. Gen. Microbiol.* 34, 483–490.

Krichevsky, M. I., and Love, L. L. (1964b). Adenine inhibition of the rate of sorocarp formation in *Dictyostelium discoideum*. *J. Gen. Microbiol.* 37, 293–295.

Krichevsky, M. I., and Love, L. L. (1965). Efflux of macromolecules from washed *Dictyostelium discoideum*. *J. Gen. Microbiol.* 41, 367–374.

Krichevsky, M. I., and Wright, B. E. (1963). Environmental control of the course of development in *Dictyostelium discoideum*. *J. Gen. Microbiol.* 32, 195–207.

Krichevsky, M. I., Love, L. L., and Chassy, B. (1969). Acceleration of morphogenesis in *D. discoideum* by exogenous mononucleotides. *J. Gen. Microbiol.* **57**, 383–389.

Krivanek, J. O. (1956). Alkaline phosphatase activity in the developing slime mold, *Dictyostelium discoideum* Raper. *J. Exp. Zool.* **133**, 459–480.

Krivanek, J. O. (1964). Nucleic acids in the developing slime mold, *Dictyostelium discoideum* (abstract). *Bull. Ass. Southeast. Biol.* **11**, 49.

Krivanek, J. O., and Krivanek, R. C. (1957). A method for embedding small specimens. *Stain Technol.* **32**, 300–301.

Krivanek, J. O., and Krivanek, R. C. (1958). The histochemical localization of certain biochemical intermediates and enzymes in the developing slime mold, *Dictyostelium discoideum* Raper. *J. Exp. Zool.* **137**, 89–115.

Krivanek, J. O., and Krivanek, R. C. (1959). Chromatographic analyses of amino acids in the developing slime mold, *Dictyostelium discoideum* Raper. *Biol. Bull.* **116**, 265–271.

Krivanek, J. O., and Krivanek, R. C. (1962). Evidence for the occurrence of transamination in the developing slime mold, *Dictyostelium discoideum* (abstract). *Amer. Zool.* **2**, 421.

Krivanek, J. O., and Krivanek, R. C. (1965). Evidence for transaminase activity in the slime mold, *Dictyostelium discoideum* Raper. *Biol. Bull.* **129**, 295–302.

Lang, A. (1954). Entwicklungsphysiologie der Acrasiales. *Fortschr. Bot.* **15**, 400–475.

Leach, C. K., and Ashworth, J. M. (1972). Characterization of DNA from the cellular slime mould *Dictyostelium discoideum* after growth of the amoebae in different media. *J. Mol. Biol.* **68**, 35–48.

Leach, C. K., Ashworth, J. M., and Garrod, D. R. (1973). Cell sorting out during the differentiation of mixtures of metabolically distinct populations of *Dictyostelium discoideum*. *J. Embryol. Exp. Morphol.* **29**, 647–661.

Leahy, D. R., McLean, E. R., Jr., and Bonner, J. T. (1970). Evidence of cyclic-3′,5′-adenosine monophosphate as chemotactic agent for polymorphonuclear leukocytes. *Blood* **36**, 52–54.

Lee, K.-C. (1972a). Cell electrophoresis of the cellular slime mould *Dictyostelium discoideum*. I. Characterization of some of the cell surface ionogenic groups. *J. Cell Sci.* **10**, 229–248.

Lee, K.-C. (1972b). Cell electrophoresis of the cellular slime mould *Dictyostelium discoideum*. II. Relevance of the changes in cell surface charge density to cell aggregation and morphogenesis. *J. Cell Sci.* **10**, 249–265.

Lee, K.-C. (1972c). Permeability of *Dictyostelium discoideum* towards amino acids and inulin: A possible relationship between initiation of differentiation, and loss of "Pool" metabolites. *J. Gen. Microbiol.* **72**, 457–471.

Lenfant, M., Ellouz, R., Das, B. C., Zissmann, E., and Lederer, E. (1969). Sur la biosynthèse de la chaîne latérale éthyle des stérols du myxomycète *Dictyostelium discoideum*. *Eur. J. Biochem.* **7**, 159–164.

Liddel, G. U., and Wright, B. E. (1961). The effect of glucose on respiration of the differentiating slime mold. *Develop. Biol.* **3**, 265–276.

Lin, S., Santi, D., and Spudich, J. (1974). Biochemical studies on the mode of action of cytochalasin B. *J. Biol. Chem.* (in press).

Lodish, H., Firtel, R., and Jacobson, A. (1973). Transcription and structure of the genome of the cellular slime mold *Dictyostelium discoideum*. *Cold Spring Harbor Symp. Quant. Biol.* **38**, 899–907.

Long, B. H., and Coe, E. L. (1973). Characterization of ubiquinone from vegetative amoebae and mature sorocarps of the cellular slime mold, *Dictyostelium discoideum*. *Comp. Biochem. Physiol.* **45B**, 933–943.

Long, B. H., and Coe, E. L. (1974). Changes in neutral lipid constituents during differentiation of the cellular slime mold *Dictyostelium discoideum*. *J. Biol. Chem.* **249**, 521–529.

Loomis, W. F., Jr. (1968). The relation between cytodifferentiation and inactivation of a developmentally-controlled enzyme in *Dictyostelium discoideum*. *Exp. Cell Res.* **53**, 282–287.

Loomis, W. F., Jr. (1969a). Acetylglucosaminidase, an early enzyme in the development of *Dictyostelium discoideum*. *J. Bacteriol.* **97**, 1149–1154.

Loomis, W. F., Jr. (1969b). Temperature-sensitive mutants of *Dictyostelium discoideum*. *J. Bacteriol.* **99**, 65–69.

Loomis, W. F., Jr. (1969c). Developmental regulation of alkaline phosphatase in *Dictyostelium discoideum*. *J. Bacteriol.* **100**, 417–422.

Loomis, W. F., Jr. (1970a). Developmental regulation of α-mannosidase in *Dictyostelium discoideum*. *J. Bacteriol.* **103**, 375–381.

Loomis, W. F., Jr. (1970b). Mutants in phototaxis of *Dictyostelium discoideum*. *Nature (London)* **227**, 745–746.

Loomis, W. F., Jr. (1970c). Temporal control of differentiation in the slime mold, *Dictyostelium discoideum*. *Exp. Cell Res.* **60**, 285–289.

Loomis, W. F., Jr. (1971). Sensitivity of *Dictyostelium discoideum* to nucleic acid analogues. *Exp. Cell Res.* **64**, 484–486.

Loomis, W. F., Jr. (1972). Role of the surface sheath in the control of morphogenesis in *Dictyostelium discoideum*. *Nature (London) New Biol.* **240**, 6–9.

Loomis, W. F., Jr., (1974). Stage specific isozymes of *Dictyostelium discoideum*. *Proc. Int. Cong. Isozymes 3rd* (C. Markert, ed.), Academic Press, New York.

Loomis, W. F., Jr., and Ashworth, J. M. (1968). Plaque-size mutants of the cellular slime mold *Dictyostelium discoideum*. *J. Gen. Microbiol.* **53**, 181–186.

Loomis, W. F., Jr., and Sussman, M. (1966). Commitment to the synthesis of a specific enzyme during cellular slime mold development. *J. Mol. Biol.* **22**, 401–404.

MacInnes, M., and Francis, D. (1974). Meiosis in *Dictyostelium mucoroides*. *Nature (London)* **251**, 321–323.

McMahon, D. (1973). A cell contact model for cellular position determination in development. *Proc. Nat. Acad. Sci. U.S.* **70**, 2396–2400.

Maeda, Y. (1970). Influence of ionic conditions on cell differentiation and morphogenesis of the cellular slime molds. *Develop. Growth Differentiat.* **12**, 217–227.

Maeda, Y. (1971). Formation of a prespore specific structure from a mitochondrion during development of the cellular slime mold *Dictyostelium discoideum*. *Develop. Growth Differentiat.* **13**, 211–219.

Maeda, Y., and Maeda, M. (1973). The calcium content of the cellular slime mold *Dictyostelium discoideum* during development and differentiation. *Exp. Cell Res.* **82**, 131–136.

Maeda, Y., and Maeda, M. (1974). Heterogeneity of the cell population of the cellular slime mold *Dictyostelium discoideum* before aggregation, and its relation to subsequent locations of the cells. *Exp. Cell Res.* **84**, 88–94.

Maeda, Y., and Takeuchi, I. (1969). Cell differentiation and fine structures in the development of the cellular slime molds. *Develop. Growth Differentiat.* **11**, 232–245.

Maeda, Y., Sugita, K., and Takeuchi, I. (1973). Fractionation of the differentiated types of cells constituting the pseudoplasmodia of the cellular slime molds. *Bot. Mag. Tokyo* **86**, 5–12.

Malchow, D., and Gerisch, G. (1973a). Cyclic AMP binding to living cells of *Dictyostelium discoideum* in presence of excess cyclic GMP. *Biochem. Biophys. Res. Commun.* **55**, 200–204.

Malchow, D., and Gerisch, G. (1973b). Recognition of extracellular cyclic AMP by aggregating cells of the slime mold *Dictyostelium discoideum*. *Hoppe-Seyler's Z. Physiol. Chem.* **354**, 1222–1223.

Malchow, D., and Gerisch, G. (1974). Short-term binding and hydrolysis of cyclic 3'5' adenosine monophosphate by aggregating *Dictyostelium* cells. *Proc. Nat. Acad. Sci. U.S.*, **71**, 2423–2427.

Malchow, D., Lüderitz, O., Westphal, O., Gerisch, G., and Riedel V. (1967). Polysaccharide in vegetativen und aggregationsreifen Amöben von *Dictyostelium discoideum*. 1. *In vivo* degradierung von bakterien-lipopoly–saccharid. *Eur. J. Biochem.* **2**, 469–479.

Malchow, D., Lüderitz, O., Kickhöfen, B., Westphal, O., and Gerisch, G. (1969). Polysaccharides in vegetative and aggregation-competent amoebae of *Dictyostelium discoideum*. 2. Purification and characterization of amoeba-degraded bacterial polysaccharides. *Eur. J. Biochem.* **7**, 239–246.

Malchow, D., Nägele, B., Schwarz, H., and Gerisch, G. (1972). Membrane-bound cyclic AMP phosphodiesterase in chemotactically responding cells of *Dictyostelium discoideum*. *Eur. J. Biochem.* **28**, 136–142.

Malchow, D., Fuchila, J., and Jastorff, B. (1973). Correlation of substrate specificity of cAMP-phosphodiesterase in *Dictyostelium discoideum* with chemotactic activity of cAMP-analogues. *FEBS Lett.* **34**, 5–9.

Malkinson, A. M., and Ashworth, J. M. (1972). Extracellular concentrations of adenosine 3'5' cyclic monophosphate during axenic growth of myxamoebae of the cellular slime mould *Dictyostelium discoideum*. *Biochem. J.* **127**, 611–612.

Malkinson, A. M., and Ashworth, J. M. (1973). Adenosine 3'5' cyclic monophosphate concentrations and phosphodiesterase activities during axenic growth and differentiation of cells of the cellular slime mould *Dictyostelium discoideum*. *Biochem. J.* **134**, 311–319.

Malkinson, A. M., Kwasniak, J., and Ashworth, J. M. (1973). Adenosine 3':5'-cyclic monophosphate–binding protein from the cellular slime mould *Dictyostelium discoideum*. *Biochem. J.* **133**, 601–603.

Marshall, R., Sargent, D., and Wright, B. E. (1970). Glycogen turnover in *Dictyostelium discoideum*. *Biochemistry*, **9**, 3087–3094.

Mason, J. W., Rasmussen, H., and DiBella, F. (1971). 3'5'AMP and Ca^{+2} in slime mold aggregation. *Exp. Cell Res.* **67**, 156–160.

Mercer, E. H., and Shaffer, B. M. (1960). Electron microscopy of solitary and aggregated slime mould cells. *J. Biophys. Biochem. Cytol.* **7**, 353–356.

Michalska, I., and Skupienski, F. X. (1939). Recherches ecologigue sur les acrasiees *Polysphondylium pallidum* Olive, *Polysphondylium violaceum* bref., *Dictyostelium mucoroides* Bref. *C. R. Acad. Sci.* **207**, 1239–1241.

Miller, Z. I., Quance, J., and Ashworth, J. M. (1969). Biochemical and cytological heterogeneity of the differentiating cells of the cellular slime mould *Dictyostelium discoideum*. *Biochem. J.* **114**, 815–818.

Mine, H., and Takeuchi, I. (1967). Tetrazolium reduction in slime mould development. *Annu. Report Biol. Works, Fac. Sci. Osaka Univ.* **15**, 97–111.

Mizukami, Y., and Iwabuchi, M. (1970a). Differential synthesis of ribosomal subunits during development in the cellular slime mold, *Dictyostelium discoideum. J. Biochem.* **67**, 501–504.

Mizukami, Y., and Iwabuchi, M. (1970b). Effects of actinomycin D and cyclohexamide on morphogenesis and syntheses of RNA and protein in the cellular slime mold, *Dictyostelium discoideum. Exp. Cell Res.* **63**, 317–324.

Mizukami, Y., and Iwabuchi, M. (1972). The formation of ribosomal subunits in the cellular slime mold *Dictyostelium discoideum. Biochim. Biophys. Acta* **272**, 81–94.

Mühlethaler, K. (1956). Electron microscopic study of the slime mold *Dictyostelium discoideum. Amer. J. Bot.* **43**, 673–678.

Muller, U. (1972). Differenzierung und musterbildung im pseudoplasmodium von *Dictyostelium discoideum.* M.S. Thesis, Univ. of Zurich.

Muller, U., and Hohl, H. R. (1973). Pattern formation in *Dictyostelium discoideum:* Temporal and spatial distribution of prespore vacuoles. *Differentiation* **1**, 267–276.

Nestle, M., and Sussman, M. (1972). The effect of cyclic AMP on morphogenesis and enzyme accumulation in *Dictyostelium discoideum. Develop. Biol.* **28**, 545–554.

Newell, P. C. (1971). The development of the cellular slime mould *Dictyostelium discoideum:* A model system for the study of cellular differentiation. *In* "Essays in Biochemistry" (P. N. Campbell and F. Dickens, eds.), Vol. 7, pp. 87–126, Academic Press, New York.

Newell, P. C. (1973). Control of development in the cellular slime mold *Dictyostelium. Biochem. Soc. Trans.* **1**, 1025.

Newell, P. C., and Sussman, M. (1969). Uridine diphosphate glucose pyrophosphorylase in *Dictyostelium discoideum.* Stability and developmental fate. *J. Biol. Chem.* **244**, 2990–2995.

Newell, P. C., and Sussman, M. (1970). Regulation of enzyme synthesis by slime mold cell assemblies embarked upon alternative developmental programmes. *J. Mol. Biol.* **49**, 627–637.

Newell, P. C., Telser, A., and Sussman, M. (1969). Alternative developmental pathways determined by environmental conditions in the cellular slime mold *Dictyostelium discoideum. J. Bacteriol.* **100**, 763–768.

Newell, P. C., Ellingson, J. S., and Sussman, M. (1969). Synchrony of enzyme accumulation in a population of differentiating slime mold cells. *Biochim. Biophys. Acta* **177**, 610–614.

Newell, P. C., Longlands, M., and Sussman, M. (1971). Control of enzyme synthesis by cellular interaction during development of the cellular slime mold *Dictyostelium discoideum. J. Mol. Biol.* **58**, 541–554.

Newell, P. C., Franke, J., and Sussman, M. (1972). Regulation of four functionally related enzymes during shifts in the developmental program of *Dictyostelium discoideum. J. Mol. Biol.* **63**, 373–382.

Nickerson, A. W., and Raper, K. B. (1973a). Macrocysts in the life cycle of the Dictyosteliaceae. I. Formation of the macrocysts. *Amer. J. Bot.* **60**, 190–197.

Nickerson, A. W., and Raper, K. B. (1973b). Macrocysts in the life cycle of the Dictyosteliaceae. II. Germination of the macrocysts. *Amer. J. Bot.* **60**, 247–254.

Nigam, V. N., Malchow, D., Rietschel, E. T., Lüderitz, O., and Westphal, O. (1970). Die enzymatische abspaltung langkettiger fettsäuren aus bakteriellen lipopolysacchariden mittels extrakten aus der amöbe von *Dictyostelium discoideum. Hoppe-Seyler's Z. Physiol. Chem.* **351**, 1123–1132.

Obata, Y., Hiroshi, A., Tanaka, Y., Yanagisawa, K., and Uchiyama, M. (1973). Isolation of a spore germination inhibitor from a cellular slime mold *Dictyostelium discoideum*. *Agr. Biol. Chem.* **37**, 1989–1990.

Ochiai, H., and Iwabuchi, M. (1971). A method for the extraction of ribosomal proteins from *Dictyostelium discoideum* with calcium chloride-urea mixture. *Bot. Mag.* **84**, 267–269.

Ochiai, H., Kanda, F., and Iwabuchi, M. (1973). The number and size of ribosomal proteins in the cellular slime mold *Dictyostelium discoideum*. *J. Biochem.* **73**, 163–167.

O'Day, D. H. (1973). Intracellular and extracellular acetylglucosaminidase activity during microcyst formation in *Polysphondylium pallidum*. *Exp. Cell Res.* **79**, 186–190.

Oehler, R. (1922). *Dictyostelium mucoroides* (Brefeld). *Zentraebl. Bakteriol. Parasitenk.* **89**, 155–156.

Olive, E. W. (1901). Preliminary enumeration of the sorophoreae. *Proc. Amer. Acad. Sci.* **37**, 333–344.

Olive, E. W. (1902). Monograph of the Acrasieae. *Proc. Boston Soc. Natur. Hist.* **30**, 451–513.

Olive, L. S. (1963). The question of sexuality in cellular slime molds. *Bull. Torrey Bot. Club* **90**, 144–147.

Olive, L. S. (1964). A new member of the Mycetozoa. *Mycologia* **61**, 885–896.

Olive, L. S. (1965). A developmental study of *Guttulinopsis vulgaris* (Acrasiales). *Amer. J. Bot.* **52**, 513–519.

Olive, L. S., and Stoianovitch, C. (1960). Two new members of the Acrasiales. *Bull. Torrey Bot. Club* **87**, 1–20.

Olive, L. S., and Stoianovitch, C. (1966). A simple new mycetozoan with ballistospores. *Amer. J. Bot.* **53**, 344–349.

Ono, H., Kobayashi, S., and Yanagisawa, K. (1972). Cell fusion in the cellular slime mold, *Dictyostelium discoideum*. *J. Cell Biol.* **54**, 665–666.

Osborn, P., and Ashworth, J. (1973). Is the ribosomal RNA synthesized during the developmental phase of the life cycle of *Dictyostelium discoideum* the same as that synthesized during the growth phase. *J. Gen. Microbiol.* **77**, 2–3.

Paddock, R. B. (1953). The appearance of amoebae tracks in cultures of *Dictyostelium discoideum*. *Science* **118**, 597–598.

Pan, P., Hall, E., and Bonner, J. T. (1972). Folic acid as secondary chemotaxic substance in the cellular slime molds. *Nature (London) New Biol.* **237**, 181–182.

Pan, P., Bonner, J. T., Wedner, H., and Parker, C. (1974). Immunofluorescence evidence for the distribution of cyclic AMP in cells and cell masses of the cellular slime molds. *Proc. Nat. Acad. Sci. U.S.* **71**, 1623–1625.

Pannbacker, R. G. (1966). RNA metabolism during differentiation in the cellular slime mold. *Biochem. Biophys. Res. Commun.* **24**, 340–345.

Pannbacker, R. G. (1967a). Uridine diphosphoglucose biosynthesis during differentiation in the cellular slime mold. I. *In vivo* measurements. *Biochemistry* **6**, 1283–1286.

Pannbacker, R. G. (1967b). Uridine diphosphoglucose biosynthesis during differentiation in the cellular slime mold. II. *In vitro* measurements. *Biochemistry* **6**, 1287–1293.

Pannbacker, R. G., and Bravard, L. J. (1972). Phosphodiesterase in *Dictyostelium discoideum* and the chemotactic response to cyclic adenosine monosphosphate. *Science* **175**, 1014–1015.

Pannbacker, R. G., and Wright, B. E. (1966). The effect of actinomycin D on development in the cellular slime mold. *Biochem. Biophys. Res. Commun.* **24,** 334–339.

Pavillard, J. (1953). Ordre des Acrasiés. *In* "Traité de Zoologie" (P. Grassé, ed.), Vol. I/II, pp. 493–505, Masson, Paris.

Payez, J., and Deering, R. (1972). Synergistic and antagonistic effects of caffeine on two strains of cellular slime molds treated with alkylating agents. *Mutat. Res.* **16,** 318–321.

Pfützner-Eckert, R. (1950). Entwicklungsphysiologische Untersuchungen an *Dictyostelium mucoroides* Brefeld. *Wilhelm Roux Entwicklungsmech. Organismen* **144,** 381–409.

Phillips, W. D., Rich, A., and Sussman, R. R. (1964). The isolation and identification of polyribosomes from cellular slime molds. *Biochim. Biophys. Acta* **80,** 508–510.

Pillinger, D., and Borek, E. (1969). Transfer RNA methylases during morphogenesis in the cellular slime mold. *Proc. Nat. Acad. Sci. U.S.* **62,** 1145–1150.

Pinoy, E. (1907). Rôle des bactéries dans le développment de certains myxomycètes. *Annu. Inst. Pasteur Paris* **21,** 622–656, **21,** 686–700.

Pinoy, P. E. (1950). Quelques observations sur la culture d'une Acrasiée. *Bull. Soc. Mycol. France* **66,** 37–38.

Poff, K., and Butler, W. (1974). Spectral characteristics of the photoreceptor pigment of phototaxis in *Dictyostelium discoideum. Nature* (*London*) **248,** 799–801.

Poff, K., and Loomis, W. F., Jr. (1973). Control of phototactic migration in *Dictyostelium discoideum. Exp. Cell Res.* **82,** 236–240.

Poff, K., Butler, W., and Loomis, W. F., Jr. (1973). Light induced absorbance changes associated with phototaxis in *Dictyostelium. Proc. Nat. Acad. Sci. U.S.* **70,** 813–816.

Poff, K., Loomis, W. F., Jr., and Butler, W. (1974). Isolation and purification of the photoreceptor pigment associated with phototaxis in *Dictyostelium discoideum. J. Biol. Chem.* **249,** 2164–2168.

Pong, S. S., and Loomis, W. F., Jr. (1971). Enzymes of amino acid metabolism in *Dictyostelium discoideum. J. Biol. Chem.* **246,** 4412–4416.

Pong, S. S., and Loomis, W. F., Jr. (1973a). Multiple nuclear RNA polymerases during development of *Dictyostelium discoideum. J. Biol. Chem.* **248,** 3933–3939.

Pong, S. S., and Loomis, W. F., Jr. (1973b). Replacement of an anabolic threonine deaminase by a catabolic threonine deaminase during development of *Dictyostelium discoideum. J. Biol. Chem.* **248,** 4867–4873.

Pong, S. S., and Loomis, W. F., Jr. (1973c). Isolation of multiple RNA polymerases from *Dictyostelium discoideum. In* "Molecular Techniques and Approaches in Developmental Biology" (M. Chrispeels, ed.), pp. 93–115, Wiley, New York.

Potts, G. (1902). Zur physiologie des *Dictyostelium mucoroides. Flora* **91,** 281–347.

Quance, J., and Ashworth, J. M. (1972). Enzyme synthesis in the cellular slime mould *Dictyostelium discoideum* during the differentiation of myxamoebae grown axenically. *Biochem. J.* **126,** 609–615.

Rafaeli, D. C. (1962). Studies on mixed morphological mutants of *Polysphondylium violaceum. Bull. Torrey Bot. Club* **89,** 312–318.

Rai, J. N., and Tewari, J. P. (1961). Studies in cellular slime moulds from Indian soils. I. On the occurrence of *Dictyostelium mucoroides* bref. and *Polysphondylium violaceum* bref. *Proc. Indian Acad. Sci.* **53,** 1–9.

Rai, J. N., and Tewari, J. P. (1963a). Studies in cellular slime moulds from Indian soils. II. On the occurrence of an aberrant strain of *Polysphondylium violaceum*

Bref., with a discussion on the relevance of mode of branching of the sorocarp as a criterion for classifying members of Dictyosteliaceae. *Proc. Indian Acad. Sci.* **58**, 201–206.

Rai, J. N., and Tewari, J. P. (1963b). Studies in cellular slime moulds from Indian soils. III. On the occurrence of two strains of *Dictyostelium mucoroides*-complex, conforming to the species *Dictyostelium sphaerocephalum* (Oud). Saccardo and March. *Proc. Indian Acad. Sci.* **58**, 263–266.

Raper, K. B. (1935). *Dictyostelium discoideum,* a new species of slime mold from decaying forest leaves. *J. Agr. Res.* **50**, 135–147.

Raper, K. B. (1937). Growth and development of *Dictyostelium discoideum* with different bacterial associates. *J. Agr. Res.* **55**, 289–316.

Raper, K. B. (1939). Influence of culture conditions upon the growth and development of *Dictyostelium discoideum. J. Agr. Res.* **58**, 157–198.

Raper, K. B. (1940a). The communal nature of the fruiting process in the acrasieae. *Amer. J. Bot.* **27**, 436–448.

Raper, K. B. (1940b). Pseudoplasmodium formation and organization in *Dictyostelium discoideum. J. E. Mitchell Sci. Soc.* **56**, 241–282.

Raper, K. B. (1941a). *Dictyostelium minutum,* a second new species of slime mold from decaying forest leaves. *Mycologia* **33**, 633–649.

Raper, K. B. (1941b). Developmental patterns in simple slime molds. Third Growth Symp. *Growth* **5**, 41–76.

Raper, K. B. (1951). Isolation, cultivation, and conservation of simple slime molds. *Quart. Rev. Biol.* **26**, 169–190.

Raper, K. B. (1956a). Factors affecting growth and differentiation in simple slime molds. *Mycologia* **48**, 169–205.

Raper, K. B. (1956b). *Dictyostelium polycephalum* n. sp.: A new cellular slime mold with coremiform fructifications. *J. Gen. Microbiol.* **14**, 716–732.

Raper, K. B. (1960a). Levels of cellular interaction in amoeboid populations. *Proc. Amer. Phil. Soc.* **104**, 579–604.

Raper, K. B. (1960b). Acrasiales. "McGraw-Hill Encyclopedia of Science and Technology" Vol. 1, p. 49–50. McGraw-Hill, New York.

Raper, K. B. (1963). The environment and morphogenesis in cellular slime molds. *Harvey Lect.* **57**, 111–141.

Raper, K. B., and Cavender, J. C. (1968). *Dictyostelium rosarium:* A new cellular slime mold with beaded sorocarps. *J. E. Mitchell Sci. Soc.* **84**, 31–47.

Raper, K. B., and Fennell, D. I. (1952). Stalk formation in *Dictyostelium. Bull. Torrey Bot. Club* **79**, 25–51.

Raper, K. B., and Fennell, D. I. (1967). The crampon-based Dictyostelia. *Amer. J. Bot.* **54**, 515–528.

Raper, K. B., and Quinlan, M. S. (1958). *Actyostelium leptosomum:* A unique cellular slime mold with an acellular stalk. *J. Gen. Microbiol.* **18**, 16–32.

Raper, K. B., and Smith, N. R. (1939). The growth of *Dictyostelium discoideum* upon pathogenic bacteria. *J. Bacteriol.* **38**, 431–444.

Raper, K. B., and Thom, C. (1932). The distribution of *Dictyostelium* and other slime molds in soil. *J. Wash. Acad. Sci.* **22**, 93–96.

Raper, K. B., and Thom, C. (1941). Interspecific mixtures in the dictyosteliaceae. *Amer. J. Bot.* **28**, 69–78.

Riedel, V., and Gerisch, G. (1968). Isolierung der Zellmembranen von kollektiven amöben (*acrasina*) mit hilfe von digitonin und filipin. *Naturwissenschaften.* **12**, 656.

Riedel, V., and Gerisch, G. (1969). Unterschiede im makromolekülbestand zwischen vegetativen und aggregationsreifen zellen von *Dictyostelium discoideum* (*acrasina*). *Wilhelm Roux Archiv. Entwicklungsmech. Organismen* 162, 268–285.

Riedel, V., and Gerisch, G. (1971). Regulation of extracellular cyclic-AMP-phosphodiesterase activity during development of *Dictyostelium discoideum*. *Biochem. Biophys. Res. Commun.* 42, 119–124.

Riedel, V., Malchow, D., Gerisch, G., and Nägele, B. (1972). Cyclic AMP phosphodiesterase interaction with its inhibitor of the slime mold, *Dictyostelium discoideum*. *Biochem. Biophys. Res. Commun.* 46, 279–287.

Riedel, V., Gerisch, G., Müller, E., and Beug, H. (1973). Defective cyclic adenosine-$3',5'$-phosphate-phosphodiesterase regulation in morphogenetic mutants of *Dictyostelium discoideum*. *J. Mol. Biol.* 74, 573–585.

Robertson, A. (1972). Quantitative analysis of the development of cellular slime molds. *Lect. Math. Life Sci.* 4, 47–73.

Robertson, A., and Cohen, M. H. (1972). Control of developing fields. *Annu. Rev. Biophys. Bioeng.* 1, 409–464.

Robertson, A., Drage, D. J., and Cohen, M. H. (1971). Control of aggregation in *Dictyostelium discoideum* by an external periodic pulse of cyclic adenosine monophosphate. *Science* 175, 333–335.

Robertson, A., Cohen, M. H., Drage, D. J., Rubin, J., Wonio, D., and Durston, A. J. (1972). Cellular interactions in slime mold aggregation. *Int. Proc. Cell Interactions* [Proc. of 3rd le Petit. Colloq., London, Nov. 1971], (L. G. Silverstri, ed.), pp. 299–306. Amer. Elsevier, New York.

Rosen, O. M., Rosen, S. M., and Horecker, B. L. (1965). Fate of the cell wall of *Salmonella typhimurium* upon ingestion by the cellular slime mold, *Polysphondylium pallidum*. *Biochem. Biophys. Res. Commun.* 18, 270–276.

Rosen, S., Kafka, J. A., Simpson, D. L., and Barondes, S. H. (1973). Developmentally-regulated, carbohydrate-binding protein in *Dictyostelium discoideum*. *Proc. Nat. Acad. Sci. U.S.* 70, 2554–2557.

Rosness, P. A. (1968). Cellulolytic enzymes during morphogenesis in *Dictyostelium discoideum*. *J. Bacteriol.* 96, 639–645.

Rosness, P. A., Gustafson, G., and Wright, B. E. (1971). Effects of adenosine $3',5'$-monophosphate and adenosine $5'$-monophosphate on glycogen degradation and synthesis in *Dictyostelium discoideum*. *J. Bacteriol.* 108, 1329–1337.

Ross, I .K. (1960). Studies on diploid strains of *Dictyostelium discoideum*. *Amer. J. Bot.* 47, 54–59.

Rossomando, E. F. (1974). Preparation of a membrane bound adenyl cyclase from *Dictyostelium discoideum* using amphitericin B. *Fed. Proc. Fed. Amer. Soc. Exp. Biol.* 33, 1362.

Rossomando, E. F., and Sussman, M. (1972). Adenyl cyclase in *Dictyostelium discoideum*: A possible control element of the chemotactic system. *Biochem. Biophys. Res. Commun.* 47, 604–610.

Rossomando, E. F., and Sussman, M. (1973). A $5'$-adenosine monophosphate–dependent adenylate cyclase and an adenosine $3':5'$-cyclic monophosphate–dependent adenosine triphosphate pyrophosphohydrolase in *Dictyostelium discoideum*. *Proc. Nat. Acad. Sci. U.S.* 70, 1254–1257.

Rossomando, E. F., Steffek, A., Mujwid, D., and Alexander, S. (1974). Scanning electron microscopic observations on cell surface changes during aggregation of *Dictyostelium discoideum*. *Exp. Cell Res.* 85, 73–78.

Roth, R., and Sussman, M. (1966). Trehalose synthesis in the cellular slime mold *Dictyostelium discoideum. Biochim. Biophys. Acta* 122, 225–231.

Roth, R., and Sussman, M. (1968). Trehalose 6-phosphate synthetase (uridine diphosphate glucose: D-glucose 6-phosphate 1-glucosyltransferase) and its regulation during slime mold development. *J. Biol. Chem.* 243, 5081–5087.

Roth, R., Ashworth, J. M., and Sussman, M. (1968). Periods of genetic transcription required for the synthesis of three enzymes during cellular slime mold development. *Proc. Nat. Acad. Sci. U.S.* 59, 1235–1242.

Runyon, E. H. (1942). Aggregation of separate cells of *Dictyostelium* to form a multicellular body. *Collecting Net* 17, 88.

Russell, G. K., and Bonner, J. T. (1960). A note on spore germination in the cellular slime mold *Dictyostelium mucoroides. Bull. Torrey Bot. Club* 87,187–191.

Rutherford, C. L., and Wright, B. E. (1971). Nucleotide metabolism during differentiation in *Dictyostelium discoideum. J. Bacteriol.* 108, 269–275.

Sackin, M. J., and Ashworth, J. M. (1969). An analysis of the distribution of volumes amongst spores of the cellular slime mould *Dictyostelium discoideum. J. Gen. Microbiol.* 59, 275–284.

Sakai, Y. (1973). Cell type conversion in isolated prestalk and prespore fragments of the cellular slime mold *Dictyostelium discoideum. Develop. Growth Differentiat.* 15, 11–19.

Sakai, Y., and Takeuchi, I. (1971). Changes of the prespore specific structure during dedifferentiation and cell type conversion of a slime mold cell. *Develop. Growth Differentiat.* 13, 231–240.

Sameshima, M., Ito, K., and Iwabuchi, M. (1972). Effect of sodium fluoride on the amount of polyribosomes, single ribosomes, and ribosomal subunits in a cellular slime mold *Dictyostelium discoideum. Biochim. Biophys. Acta* 28, 79–85.

Samuel, E. W. (1961). Orientation and rate of locomotion of individual amebas in the life cycle of the cellular slime mold *Dictyostelium mucoroides. Develop. Biol.* 3, 317–335.

Sargent, D., and Wright, B. E. (1971). Trehalose synthesis during differentiation in *Dictyostelium discoideum.* II. *In vivo* determinations. *J. Biol. Chem.* 246, 5340–5344.

Schildkraut, C. L., Mandel, M., Levisohn, S., Smith-Sonneborn, J. E., and Marmur, J. (1962). Deoxyribonucleic acid base composition and taxonomy of some protozoa. *Nature (London)* 196, 795–796.

Schuckmann, W. von (1924). Zur biologie von *Dictyostelium mucoroides* bref. *Zentralbl. Bakteriol. Parasitenk.* 91, 302–309.

Schuckmann, W. von (1925). Zur morphologie und biologie von *Dictyostelium mucoroides* Bref. *Arch. Protistenk.* 51, 495–529.

Schwalb, R., and Roth, R. (1970). Axenic growth and development of the ˙cellular slime mold *Dictyostelium discoideum. J. Gen. Microbiol.* 60, 283–286.

Segel, L., and Stoeckly, B. (1972). Instability of a layer of chemotactic cells, attractant and degrading enzyme. *J. Theor. Biol.* 37, 561–585.

Shaffer, B. M. (1953). Aggregation in cellular slime moulds: *In vitro* isolation of acrasin. *Nature (London)* 171, 975.

Shaffer, B. M. (1956a). Properties of acrasin. *Science* 123, 1172–1173.

Shaffer, B. M. (1956b). Acrasin, the chemotactic agent in cellular slime moulds. *J. Exp. Biol.* 33, 645–657.

Shaffer, B. M. (1957a). Aspects of aggregation in cellular slime molds. I. Orientation and chemotaxis. *Amer. Natur.* 91, 19–35.

Shaffer, B. M. (1957b). Properties of slime-mould amoebae of significance for aggregation. *Quart. J. Microscop. Sci.* **98**, 377–392.

Shaffer, B. M. (1957c). Variability of behaviour of aggregating cellular slime moulds. *Quart. J. Microscop. Sci.* **98**, 393–405.

Shaffer, B. M. (1957). Properties of slime mould amoebae of significance for aggregation. *Quart. J. Microscop. Sci.* **98**, 377–392.

Shaffer, B. M. (1958). Integration in aggregating cellular slime moulds. *Quart. J. Microscop. Sci.* **99**, 103–121.

Shaffer, B. M. (1961a). Species differences in the aggregation of the Acrasieae *In* "Recent Advances in Botany," *Proc. 9th Int. Botan. Congr.,* pp. 294–298. Univ. of Toronto Press, Toronto.

Shaffer, B. M. (1961b). The cells founding aggregation centres in the slime mould *Polysphondylium violaceum. J. Exp. Biol.* **38**, 833–849.

Shaffer, B. M. (1962). The Acrasina. *Advan. Morphog.* **2**, 109–182.

Shaffer, B. M. (1963a). Inhibition by existing aggregations of founder differentiation in the cellular slime mould *Polysphondylium violaceum. Exp. Cell Res.* **31**, 432–535.

Shaffer, B. M. (1963b). Behaviour of particles adhering to amoebae of the slime mould *Polysphondylium violaceum* and the fate of the cell surface during locomotion. *Exp. Cell Res.* **32**, 603–606.

Shaffer, B. M. (1964a). Intracellular movement and locomotion of cellular slime mold amoebae. *In* "Primitive Motile Systems in Cell Biology" (R. D. Allen and N. Kaniva, eds.), pp. 387–405. Academic Press, New York.

Shaffer, B. M. (1964b). Attraction through air exerted by unaggregated cells on aggregates of the slime mould *Polysphondylium violaceum. J. Gen. Microbiol.* **36**, 359–364.

Shaffer, B. M. (1964c). The Acrasina. *Advan. Morphog.* **3**, 301–322.

Shaffer, B. M. (1965a). Antistrophic pseudopodia of the collective amoeba *Polysphondylium violaceum. Exp. Cell Res.* **37**, 79–92.

Shaffer, B. M. (1965b). Mechanical control of the manufacture and resorption of cell surface in collective amoebae. *J. Theoret. Biol.* **8**, 27–40.

Shaffer, B. M. (1965c). Cell movement within aggregates of the slime mould *Dictyostelium discoideum* revealed by surface markers. *J. Embryol. Exp. Morphol.* **13**, 97–117.

Shaffer, B. M. (1965d). Pseudopodia and intracytoplasmic displacements of the collective amoebae Dictyosteliidae. *Exp. Cell Res.* **37**, 12–25.

Sharma, O. K., and Borek, E. (1970). Inhibitor of transfer ribonucleic acid methylases in the differentiating slime mold *Dictyostelium discoideum. J. Bacteriol.* **101**, 705–708.

Singh, B. N. (1946). Soil Acraieae and their bacterial food supply. *Nature (London)* **157**, 133–134.

Singh, B. N. (1947a). Studies on soil Acrasieae: I. Distribution of species of *Dictyostelium* in soils of Great Britain and the effect of bacteria on their development. *J. Gen. Microbiol.* **1**, 11–21.

Singh, B. N. (1947b). Studies on soil Acrasieae: II. The active life of species of *Dictyostelium* in soil and the influence thereon of soil moisture and bacterial food. *J. Gen. Microbiol.* **1**, 361–367.

Sinha, U., and Ashworth, J. M. (1969). Evidence for the existence of elements of a para-sexual cycle in the cellular slime mould, *Dictyostelium discoideum. Proc. Roy. Soc. Edinburgh Sect. B.* **173**, 531–540.

Skupienski, F. X. (1919). Sur la sexualité chez une espèce de Myxomycète Acrasiée, *Dictyostelium mucoroides. C. R. Acad. Sci.* **167**, 960–962.

Skupienski, F. X. (1920). "Recherches sur le cycle évolutif de certains myxomycètes." Paris. 81 pp.

Slifkin, M. K., and Bonner, J. T. (1952). The effects of salts and organic solutes on the migration of the slime mold *Dictyostelium discoideum. Biol. Bull.* **102**, 273–277.

Slifkin, M. K., and Gutowsky, H. S. (1958). Infrared spectroscopy as a new method for assessing the nutritional requirements of the slime mold, *Dictyostelium discoideum. J. Cell. Comp. Physiol.* **51**, 249–257.

Snyder, H. M., and Ceccarini, C. (1966). Interspecific spore inhibition in the cellular slime molds. *Nature (London)* **209**, 1152.

Soll, D., and Sussman, M. (1973). Transcription in isolated nuclei of the slime mold *Dictyostelium discoideum. Biochim. Biophys. Acta* **319**, 312–322.

Solomon, E. P., Johnson, E. M., and Gregg, J. H. (1964). Multiple forms of enzymes in a cellular slime mold during morphogenesis. *Develop. Biol.* **9**, 314–326.

Sonneborn, D. R., White, G. J., and Sussman, M. (1963). A mutation affecting both rate and pattern of morphogenesis in *Dictyostelium discoideum. Develop. Biol.* **7**, 79–93.

Sonneborn, D. R., Sussman, M., and Levine, L. (1964). Serological analyses of cellular slime-mold development. I. Changes in antigenic activity during cell aggregation. *J. Bacteriol.* **87**, 1321–1329.

Sonneborn, D. R., Levine, L., and Sussman, M. (1965). Serological analyses of cellular slime mold development. II. Preferential loss, during morphogenesis, of antigenic activity associated with the vegetative myxamoebae. *J. Bacteriol.* **89**, 1092–1096.

Spudich, J. (1974). Biochemical and structural studies of actomyosin-like proteins from non-muscle cells. II. Pruification, properties and membrane association of actin from amoebae of *Dictyostelium discoideum. J. Biol. Chem.* **249**, 6013–6020.

Staples, S. O., and Gregg, J. H. (1967). Carotenoid pigments in the cellular slime mold, *Dictyostelium discoideum. Biol. Bull.* **132**, 413–422.

Strong, C. L. (1966). How to cultivate the slime molds and perform experiments on them. *Sci. Amer.* **214**, 116–121.

Stuchell, R., Weinstein, B., and Beattie, D. (1973). Effects of ethidium bromide on various segments of the respiratory chain in the cellular slime mold *Dictyostelium discoideum. FEBS Lett.* **37**, 23–26.

Sussman, M. (1951). The origin of cellular heterogeneity in the slime molds, Dictyosteliaceae. *J. Exp. Zool.* **118**, 407–418.

Sussman, M. (1952). An analysis of the aggregation stage in the development of the slime molds, Dictyosteliaceae. II. Aggregative center formation by mixtures of *Dictyostelium discoideum* wild type and aggregateless variants. *Biol. Bull.* **103**, 446–457.

Sussman, M. (1954). Synergistic and antagonistic interactions between morphogenetically deficient variants of the slime mould *Dictyostelium discoideum. J. Gen. Microbiol.* **10**, 110–120.

Sussman, M. (1955a). "Fruity" and other mutants of the cellular slime mould, *Dictyostelium discoideum:* A study of developmental aberrations. *J. Gen. Microbiol.* **13**. 295–309.

Sussman, M. (1955b). The developmental physiology of the amoeboid slime molds. *In* "Biochemistry and Physiology of the Protozoa" (S. Hunter, and A. Lwoff, eds.), Vol. 2, pp. 201–223. Academic Press, New York.

Sussman, M. (1956a). On the relation between growth and morphogenesis in the slime mold *Dictyostelium discoideum. Biol. Bull.* 110, 91–95.

Sussman, M. (1956b). The biology of the cellular slime molds. *Annu. Rev. Microbiol.* 10, 21–50.

Sussman, M. (1958). A developmental analysis of cellular slime mold aggregation. *In* "A Symposium on the Chemical Basis of Development" (W. D. McElroy and B. Glass, eds.), pp. 264–295. Johns Hopkins Press, Baltimore, Maryland.

Sussman, M. (1961a). Cultivation and serial transfer of the slime mould, *Dictyostelium discoideum* in liquid nutrient medium. *J. Gen Microbiol.* 25, 375–378.

Sussman, M. (1961b). Cellular differentiation in the slime mold. *In* "Growth in Living Systems" (M. X. Zarrow, ed.), pp. 221–239. Basic Books, New York.

Sussman, M. (1963). Growth of the cellular slime mold *Polysphondylium pallidum* in a simple nutrient medium. *Science* 139, 338.

Sussman, M. (1965a). Inhibition by actidione of protein synthesis and UDP-Gal polysaccharide transferase accumulation in *Dictyostelium discoideum. Biochem. Biophys. Res. Commun.* 18, 763–767.

Sussman, M. (1965b). Temporal, spatial, and quantitative control of enzyme activity during slime mold cytodifferentiation. *In* "Genetic Control of Differentiation," pp. 66–76. Brookhaven Symposia in Biology No. 18, Brookhaven, New York.

Sussman, M. (1965c). Developmental phenomena in microorganisms and in higher forms of life. *Annu. Rev. Microbiol.* 19, 59–78.

Sussman, M. (1966a). Biochemical and genetic methods in the study of cellular slime mold development. *In* "Methods in Cell Physiology" (D. Prescott, ed.), Vol. 2, pp. 397–410. Academic Press, New York.

Sussman, M. (1966b). Protein synthesis and the temporal control of genetic transcription during slime mold development. *Proc. Nat. Acad. Sci. U.S.* 55, 813–818.

Sussman, M. (1967). Evidence for temporal and quantitative control of genetic transcription and translation during slime mold development. *Fed. Proc. Fed. Amer. Soc. Exp. Biol.* 26, 77–83.

Sussman, M. (1970). Model for quantitative and qualitative control of mRNA translation in eukaryotes. *Nature (London)* 225, 1245–1246.

Sussman, M. (1972). The program of polysaccharide and disaccharide synthesis during the development of *Dictyostelium discoideum. In* "Biochemistry of the Glycosidic Linkage" (R. Piras and H. G. Pontis, eds.), pp. 431–448. Academic Press, New York.

Sussman, M., and Bradley, S. G. (1954). A protein growth factor of bacterial origin required by the cellular slime molds. *Arch. Biochem. Biophys.* 51, 428–435.

Sussman, M., and Ennis, H. L. (1959). The role of the initiator cell in slime mold aggregation. *Biol. Bull.* 116, 304–317.

Sussman, M., and Lee, F. (1954). Physiology of developmental variants among the cellular slime molds (abstract). *Bacteriol. Proc.,* 42.

Sussman, M., and Lee, F. (1955). Interactions among variant and wild-type strains of cellular slime molds across thin agar membranes. *Proc. Nat. Acad. Sci. U.S.* 41, 70–78.

Sussman, M., and Lovgren, N. (1965). Preferential release of the enzyme UDP-galactose polysaccharide transferase during cellular differentiation in the slime mold, *Dictyostelium discoideum. Exp. Cell Res.* 38, 97–105.

Sussman, M., and Newell, P. C. (1972). Quantal control. *In* "Molecular Genetics and Developmental Biology" (M. Sussman, ed.), pp. 275–302. Prentice-Hall, Englewood Cliffs, New Jersey.

Sussman, M., and Noël, E. (1952). An analysis of the aggregation stage in the development of the slime molds, Dictyosteliaceae. I. The populational distribution of the capacity to initiate aggregation. *Biol. Bull.* 103, 259–268.

Sussman, M., and Osborn, M. J. (1964). UDP-galactose polysaccharide transferase in the cellular slime mold, *Dictyostelium discoideum:* Appearance and disappearance of activity during cell differentiation. *Proc. Nat. Acad. Sci. U.S.* 52, 81–87.

Sussman, M., and Sussman, R. R. (1956). Cellular interactions during the development of the cellular slime molds. *In* "Cellular Mechanisms in Differentiation and Growth" 14th Growth Symposium (D. Rudnick, ed.), pp. 125–154. Princeton University Press, Princeton, New Jersey.

Sussman, M., and Sussman, R. R. (1961). Aggregative performance. *Exp. Cell Res. Suppl.* 8, 91–106.

Sussman, M., and Sussman, R. R. (1962). Ploidal inheritance in *Dictyostelium discoideum.* I. Stable haploid, stable diploid and metastable strains. *J. Gen. Microbiol.* 28, 417–429.

Sussman, M., and Sussman, R. R. (1965). The regulatory program for UDP-Gal polysaccharide transferase activity during slime mold cytodifferentiation: Requirement for specific synthesis of RNA. *Biochim. Biophys. Acta* 108, 463–473.

Sussman, M., and Sussman, R. R. (1969). Patterns of RNA synthesis and of enzyme accumulation and disappearance during cellular slime mould cytodifferentiation. *Symp. Soc. Gen. Microbiol.* 19, 403–435.

Sussman, M., Lee, F., and Kerr, N. S. (1956). Fractionation of acrasin, a specific chemotactic agent for slime mold aggregation. *Science* 123, 1171–1172.

Sussman, M., Loomis, W. F., Jr., Ashworth, J. M., and Sussman, R. R. (1967). The effect of actinomycin D on cellular slime mold morphogenesis. *Biochem. Biophys. Res. Commun.* 26, 353–359.

Sussman, R. R. (1961). A method for staining chromosomes of *D. discoideum* myxamoebae in the vegetative stage. *Exp. Cell Res.* 24, 154–155.

Sussman, R. R. (1967). RNA metabolism during cytodifferentiation in the cellular slime mold *Polysphondylium pallidum. Biochim. Biophys. Acta* 149, 407–421.

Sussman, R. R. (1974). Bioassay for the isolation of *Dictyostelium discoideum* mutants deficient in extracellular accumulation of cyclic AMP. *J. Bacteriol.* 118, 312–313.

Sussman, R. R., and Rayner, E. P. (1971). Physical characterization of deoxyribonucleic acids in *Dictyostelium discoideum. Arch. Biochem. Biophys.* 144, 127–137.

Sussman, R. R., and Sussman, M. (1953). Cellular differentiation in Dictyosteliaceae: heritable modifications of the developmental pattern. *Ann. N.Y. Acad. Sci.* 56, 949–960.

Sussman, R. R., and Sussman, M. (1960). The dissociation of morphogenesis from cell division in the cellular slime mould, *Dictyostelium discoideum. J. Gen. Microbiol.* 23, 287–293.

Sussman, R. R., and Sussman, M. (1963). Ploidal inheritance in the slime mould *Dictyostelium discoideum:* Haploidization and genetic segregation of diploid strains. *J. Gen. Microbiol.* 30, 349–355.

Sussman, R. R., and Sussman, M. (1967). Cultivation of *Dictyostelium discoideum* in axenic medium. *Biochem. Biophys. Res. Commun.* 29, 53–55.

Sussman, R. R., Sussman, M., and Fu, F. L. (1958). The chemotactic complex responsible for cellular slime mold aggregation (abstract). *Bacteriol. Proc.* 32.

Sussman, R. R., Sussman, M., and Ennis, H. L. (1960). Appearance and inheritance of the I-cell phenotype in *D. discoideum. Develop. Biol.* **2,** 367–392.

Takeuchi, I. (1960). The correlation of cellular changes with succinic dehydrogenase and cytochrome oxidase activities in the development of the cellular slime molds. *Develop. Biol.* **2,** 343–366.

Takeuchi, I. (1963). Immunochemical and immunohistochemical studies on the development of the cellular slime mold *Dictyostelium mucoroides. Develop. Biol.* **8,** 1–26.

Takeuchi, I. (1969). Establishment of polar organization during slime mold development. *In* "Nucleic Acid Metabolism Cell Differentiation and Cancer Growth" (E. V. Cowdry, and S. Seno, eds.), pp. 297–304. Pergamon Press, Oxford.

Takeuchi, I. (1972). Differentiation and dedifferentiation in cellular slime molds. *Aspects Cell. Mol. Phys.,* 217–236.

Takeuchi, I., and Sakai, Y. (1971). Dedifferentiation of the disaggregated slug cell of the cellular slime mold *Dictyostelium discoideum. Develop. Growth Differentiat.* **13,** 201–210.

Takeuchi, I., and Sato, T. (1965). Cell differentiation and cell sorting in the development of cellular slime molds. *Jap. J. Exp. Morphol.* **19,** 67–70.

Takeuchi, I., and Tazawa, M. (1955). Studies on the morphogenesis of the slime mould, *Dictyostelium discoideum. Cytologia* **20,** 157–165.

Takeuchi, I., and Yabuno, K. (1970). Disaggregation of slime mold pseudoplasmodia using EDTA and various proteolytic enzymes. *Exp. Cell Res.* **61,** 183–190.

Tanaka, Y., Yanagisawa, K., Hashimoto, Y., and Yamaguchi, M. (1974). True spore germination inhibitor of a cellular slime mold *Dictyostelium discoideum. Agr. Biol. Chem.* **38,** 689–690.

Telser, A., and Sussman, M. (1971). Uridine diphosphate galactose-4-epimerase, a developmentally regulated enzyme in the cellular slime mold *Dictyostelium discoideum. J. Biol. Chem.* **246,** 2252–2257.

Thom, C., and Raper, K. B. (1930). Myxamoebae in soil and decomposing crop residues. *J. Wash. Acad. Sci.* **20,** 362–370.

Tieghem, P. van (1880). Sur quelques Myxomycètes à plasmode agrégé. *Bull. Soc. Bot. Fr.* **27,** 317–322.

Tieghem, P. van (1884). *Coenonia,* genre nouveau de Myxomycètes à plasmode agrégé. *Bull. Soc. Bot. Fr.* **31,** 303–306.

Toama, M. A., and Raper, K. B. (1967a). Microcysts of the cellular slime mold *Polysphondylium pallidum.* I. Factors influencing microcyst formation. *J. Bacteriol.* **94,** 1143–1149.

Toama, M. A., and Raper, K. B. (1967b). Microcysts of the cellular slime mold *Polysphondylium pallidum.* II. Chemistry of the microcyst walls. *J. Bacteriol.* **94,** 1150–1153.

Tuchman, J., Alton, T., and Lodish, H. (1974). Preferential synthesis of actin during early development of the slime mold *Dictyostelium discoideum. Develop. Biol.* **40,** 116–129.

Vuillemin, P. (1903). Une Acrasiée bactériophage. *C. R. Acad. Sci.* **137,** 387–389.

Ward, C., and Wright, B. E. (1965). Cell wall synthesis in *Dictyostelium discoideum.* I. *In vitro* synthesis from uridine diphosphoglucose. *Biochemistry* **4,** 2021–2027.

Ward, J. M. (1958). Biochemical systems involved in differentiation of the fungi. *In* "Fourth International Congress of Biochemistry," Vol. 6, pp. 33–64, (Biochemistry of Morphogenesis), Pergamon Press, Oxford.

Watts, D. J., and Ashworth, J. M. (1970). Growth of myxamoebae of the cellular slime mould *Dictyostelium discoideum* in axenic culture. *Biochem. J.* **119**, 171–174.

Weber, A. T., and Raper, K. B. (1971). Induction of fruiting in two aggregateless mutants of *Dictyostelium discoideum*. *Develop. Biol.* **26**, 606–615.

Weeks, G. (1973). Agglutination of growing and differentiating cells of *Dictyostelium discoideum* by concanavalin A. *Exp. Cell Res.* **76**, 467–470.

Weeks, G., and Ashworth, J. M. (1972). Glycogen synthetase and the control of glycogen synthesis in the cellular slime mould *Dictyostelium discoideum* during the growth (myxamoebal) phase. *Biochem. J.* **126**, 617–626.

Weiner, E., and Ashworth, J. M. (1970). The isolation and characterization of lysosomal particles from myxamoebae of the cellular slime mould *Dictyostelium discoideum*. *Biochem. J.* **118**, 505–512.

Weinkauff, A. M., and Filosa, M. F. (1965). Factors involved in the formation of macrocysts by the cellular slime mold, *Dictyostelium mucoroides*. *Can. J. Microbiol.* **11**, 385–387.

Weinstein, B., and Koritz, S. (1973). A protein kinase assayable with intact cells of the cellular slime mold *Dictyostelium discoideum*. *Develop. Biol.* **34**, 159–162.

White, G. J., and Sussman, M. (1961). Metabolism of major cell constituents during slime mold morphogenesis. *Biochim. Biophys. Acta* **53**, 285–293.

White, G. J., and Sussman, M. (1963a). Polysaccharides involved in slime-mold development. I. Water-soluble glucose polymer(s). *Biochim. Biophys. Acta* **74**, 173–178.

White, G. J., and Sussman, M. (1963b). Polysaccharides involved in slime mold development. II. Water-soluble acid mucopolysaccharide(s). *Biochim. Biophys. Acta* **74**, 179–187.

Whitfield, F. E. (1964). The use of proteolytic and other enzymes in the separation of slime mould grex. *Exp. Cell Res.* **36**, 62–72.

Williams, K., Kessin, R., and Newell, P. (1974). Genetics of growth in axenic medium of the cellular slime mould *Dictyostelium discoideum*. *Nature (London)* **247**, 142–143.

Wittingham, W. F., and Raper, K. B. (1956). Inhibition of normal pigment synthesis in spores of *Dictyostelium purpureum*. *Amer. J. Bot.* **43**, 703–708.

Wittingham, W. F., and Raper, K. B. (1957). Environmental factors influencing the growth and fructification of *Dictyostelium polycephalum*. *Amer. J. Bot.* **44**, 619–627.

Wittingham, W. F., and Raper, K. B. (1960). Non-viability of stalk cells in *Dictyostelium*. *Proc. Nat. Acad. Sci. U.S.* **46**, 642–649.

Wilson, C. M. (1952). Sexuality in the Acrasiales. *Proc. Nat. Acad. Sci. U.S.* **38**, 659–662.

Wilson, C. M. (1953). Cytological study of the life cycle of *Dictyostelium*. *Amer. J. Bot.* **40**, 714–718.

Wilson, C. M., and Ross, I. K. (1957). Further cytological studies in the Acrasiales. *Amer. J. Bot.* **44**, 345–350.

Wolpert, L. (1971). Positional information and pattern formation. *Curr. Top. Dev. Biol.* **6**, 183–222.

Woolley, D. (1970). Extraction of an actomyosin-like protein from amoeba *Dictyostelium discoideum*. *J. Cell Physiol.* **76**, 185–190.

Woolley, D. E. (1972). An actin-like protein from amoebae of *Dictyostelium discoideum*. *Arch. Biochem. Biophys.* **150**, 519–530.

Wright, B. E. (1958). Effect of steroids on aggregation in the slime mold *Dictyostelium discoideum* (abstract). *Bacteriol. Proc.*, **115.**

Wright, B. E. (1960). On enzyme-substrate relationships during biochemical differentiation. *Proc. Nat. Acad. Sci. U.S.* **46,** 798–803.

Wright, B. E. (1963a). Endogenous substrate control in biochemical differentiation. *Bacteriol. Rev.* **27,** 273–281.

Wright, B. E. (1963b). Endogenous activity and sporulation in slime molds. *Ann. N.Y. Acad. Sci.* **102,** 740–754.

Wright, B. E. (1964). Biochemistry of Acrasiales. *In* "Biochemistry and Physiology of the Protozoa" (S. H. Hutner, ed.), Vol. 3, pp. 341–381. Academic Press, New York.

Wright, B. E. (1965). Control of carbohydrate synthesis in the slime mold. *In* "Development and Metabolic Control and Neoplasia," pp. 298–316. Williams & Wilkins, Baltimore, Maryland.

Wright, B. E. (1966). Multiple causes and controls in differentiation. *Science* **153,** 830–837.

Wright, B. E. (1968). An analysis of metabolism underlying differentiation in *Dictyostelium discoideum. J. Cell Physiol. Suppl. 1,* **72,** 146–160.

Wright, B. E. (1972). Actinomycin D and genetic transcription during differentiation. *Develop. Biol.* **28,** f–13–20.

Wright, B. E. (1973). "Critical Variables in Differentiation," Prentice-Hall, Englewood Cliffs, New Jersey.

Wright, B. E., and Anderson, M. L. (1958). Enzyme patterns during differentiation in the slime mold. *In* "A Symposium on the Chemical Basis of Development" (W. D. McElroy and B. Glass, eds.), pp. 296–314. Johns Hopkins Press, Baltimore, Maryland.

Wright, B. E., and Anderson, M. L. (1959). Biochemical differentiation in the slime mold. *Biochim. Biophys. Acta* **31,** 310–322.

Wright, B. E., and Anderson, M. L. (1960a). Protein and amino acid turnover during differentiation in the slime mold. I. Utilization of endogenous amino acids and proteins. *Biochim. Biophys. Acta* **43,** 62–66.

Wright, B. E., and Anderson, M. L. (1960b). Protein and amino acid turnover during differentiation in the slime mold. II. Incorporation of 35 S methionine into the amino acid pool and into protein. *Biochim. Biophys. Acta* **43,** 67–78.

Wright, B. E., and Bard, S. (1963). Glutamate oxidation in the differentiating slime mold. I. Studies *in vivo. Biochim. Biophys. Acta* **71,** 45–49.

Wright, B. E., and Bloom, B. (1960). *In vivo* investigations of glucose catabolism in a differentiating slime mold (abstract). *Bacteriol. Proc.* **59.**

Wright, B. E., and Bloom, B. (1961). *In vivo* evidence for metabolic shifts in the differentiating slime mold. *Biochim. Biophys. Acta* **48,** 342–346.

Wright, B. E., and Brühmüller, M. (1964). The effect of exogenous glucose concentration of C-6/C-I ratio. *Biochim. Biophys. Acta* **82,** 203–204.

Wright, B. E., and Dahlberg, D. (1967). Cell wall synthesis in *Dictyostelium discoideum.* II. Synthesis of soluble glycogen by a cytoplasmic enzyme. *Biochemistry* **6,** 2074–2079.

Wright, B. E., and Dahlberg, D. (1968). Stability *in vitro* of uridine disphosphoglucose pyrophosphorylase in *Dictyostelium discoideum. J. Bacteriol.* **95,** 983–985.

Wright, B. E., and Gustafson, G. L. (1972). Expansion of the kinetic model of differentiation in *Dictyostelium discoideum. J. Biol. Chem.* **247,** 7875–7884.

Wright, B. E., and Marshall, R. (1971). Trehalose synthesis during differentiation in *Dictyostelium discoideum*. I. Analysis and predictions by computer simulation. *J. Biol. Chem.* **246**, 5335–5339.

Wright, B. E., and Pannbacker, R. (1967). Inhibition by actinomycin D of uridine diphosphoglucose synthetase activity during differentiation of *Dictyostelium discoideum*. *J. Bacteriol.* **93**, 1762–1764.

Wright, B. E., and Wassarman, M. E. (1964). Pyridine nucleotide levels in *Dictyostelium discoideum* during differentiation. *Biochim. Biophys. Acta* **90**, 423–424.

Wright, B. E., Brühmüller, M., and Ward, C. (1964). Studies *in vivo* on hexose metabolism in *Dictyostelium discoideum*. *Develop. Biol.* **9**, 287–297.

Wright, B. E., Ward, C., and Dahlberg, D. (1966). Cell wall polysaccharide synthesis *in vitro* catalyzed by an enzyme from slime mole myxamoebae lacking a cell wall. *Biochem. Biophys. Res. Commun.* **22**, 352–356.

Wright, B. E., Simon, W., and Walsh, B. T. (1968). A kinetic model of metabolism essential to differentiation in *Dictyostelium discoideum*. *Proc. Nat. Acad. Sci. U.S.* **60**, 644–651.

Wright, B. E., Dahlberg, D., and Ward, C. (1968). Cell wall synthesis in *Dictyostelium discoideum*. A model system for the synthesis of alkali-insoluble cell wall glycogen during differentiation. *Arch. Biochem. Biophys.* **124**, 380–385.

Wright, B. E., Rosness, P., Jones, T. H. D., and Marshall, R. (1973). Glycogen metabolism during differentiation in *Dictyostelium discoideum*. *Ann. N.Y. Acad. Sci.* **210**, 51–63.

Yabuno, K. (1970). Changes in electronegativity of the cell surface during the development of the cellular slime mold *Dictyostelium discoideum*. *Develop. Growth Differentiat.* **12**, 229–239.

Yabuno, Y. Y. (1971). Changes in cellular adhesiveness during the development of the slime mould *Dictyostelium discoideum*. *Develop. Growth Differentiat.* **13**, 181.

Yamada, T., Yanagisawa, K., and Sinata, Y. (1972). Inhibition of differentiation in a sporeless mutant KS 17 of *Dictyostelium discoideum*. *Cytologia* **37**, 383–388.

Yamada, T., Yanagisawa, K. O., Ono, H., and Yanagisawa, K. (1973). Genetic analysis of developmental stages of the cellular slime mold *Dictyostelium purpureum*. *Proc. Nat. Acad. Sci. U.S.* **70**, 2003–2005.

Yamada, H., Yadama, T., and Miyazaki, T. (1974). Polysaccharides of the cellular slime mold. I. Extracellular polysaccharides in growth phase of *Dictyostelium discoideum*. *Biochim. Biophys. Acta* **343**, 371–377.

Yanagida, M., and Noda, H. (1967). Cell contact and cell surface properties in the cellular slime mold, *Dictyostelium discoideum*. *Exp. Cell Res.* **45**, 399–414.

Yanagisawa, K., Loomis, W. F., Jr., and Sussman, M. (1967). Developmental regulation of the enzyme UDP-galactose polysaccharide transferase. *Exp. Cell Res.* **46**, 328–334.

Yanagisawa, K., Yamada, T., and Ono, H. (1969). A study of differentiation in the cellular slime mold by using developmental mutants. *Zool. Mag.* **78**, 277–286.

Yarger, J., Stults, K., and Soll, D. (1974). Observations on the growth of *Dictyostelium discoideum* in axenic medium: Evidence for an extracellular inhibitor synthesized by stationary phase cells. *J. Cell Sci.* **14**, 681–690.

Subject Index

A 5
B 6
C 7
D 8
E 9
F 0
G 1
H 2
I 3
J 4